Pamela A. Chambers.

CAMBRIDGE ELEMENTARY STATISTICAL TABLES

BY

D. V. LINDLEY

Professor of Statistics
University College London

AND

J. C. P. MILLER

Computer Laboratory, Cambridge

CAMBRIDGE UNIVERSITY PRESS

CAMBRIDGE

LONDON · NEW YORK · MELBOURNE

CONTENTS

CONSTANTS

$e = 2\cdot71828\ 18285$ $\qquad$ $\log_{10} e = 0\cdot43429\ 44819$

$\pi = 3\cdot14159\ 26536$ $\qquad$ $\log_e 10 = 2\cdot30258\ 50930$

$\frac{1}{\sqrt{2\pi}} = 0\cdot39894\ 22804$ $\qquad$ $\log_e \sqrt{2\pi} = 0\cdot91893\ 85332$

PREFACE

This set of tables is concerned only with the commoner and more familiar and elementary of the many statistical functions and tests of significance now available. It is hoped that the values provided will meet the majority of the needs of many users of statistical methods in scientific research, technology and industry in a compact and handy form, and that the collection will provide a convenient set of tables for the teaching and study of statistics in schools and universities.

The more familiar statistical tests are either based directly on the normal distribution or, in the case of the t-, χ^2- and F-tests, they are derived therefrom. Percentage points for these tests are provided in the tables, mainly for significance levels of 5 %, 1 % and 0·1 % in both one-sided and two-sided tests. Again, many situations where tests based on the normal distribution are not directly applicable may be studied by transforming the data so that at least approximate normal distributions obtain. Accordingly tables of the more common transformations, square root, logarithm, inverse circular and hyperbolic root-sines, together with that for the correlation coefficient, have been included.

Finally, besides tables of percentage points and transforms, the statistician needs certain functions of the integers in order to calculate the t-, χ^2- and F-criteria and tables of these functions form the greater part of the booklet.

Table 1 gives a full tabulation of the normal distribution and a brief tabulation of the normal frequency function. The distribution function, that is the area from minus infinity up to a given point, has been chosen in preference to a two-sided tabulation. Inverse interpolation in this table gives normal equivalent deviates. Tabulation of the distribution function for large values of the argument is most satisfactorily done through a critical table, and though such tables are perhaps rather unfamiliar, they are too useful and compact to be excluded on that account. Table 2 gives the percentage points (one-tailed) of the normal distribution function for selected values of the argument: the last column gives the percentage points corresponding to the usual significance levels, both one- and two-tailed.

Tables 3, 5 and 7 give the percentage points of the t-, χ^2- and F-distributions, respectively. It is important to notice that whilst χ^2 and F are given for single-tailed or one-sided tests, the values of t are for two-tailed or two-sided tests, leading to a rather different choice of percentage points. We would have preferred to have given the values of t for a one-tailed test but it is so well established in its present form that a departure from tradition might well cause more trouble than assistance. In most tables the entries are believed to be correct to the last figure given; in Table 7, a table of triple entry, there are some cases where the last figure given may be incorrect by one unit. Every effort has been made to ensure that no larger error is present.

Table 4 gives the transformation required to make correlation studies possible by normal distribution methods. Table 6 provides a convenient and rapid means of estimating the standard deviation in small samples. Table 8 consists of random numbers for use in the design of experiments and for instructional purposes in making up random samples. Tables 1 and 8 together provide a means of constructing random samples from a normal population.

Table 9 gives, in effect, certain functions of the integers up to 1000, namely the square, square root, reciprocal, reciprocal square root, logarithm and antilogarithm to the base 10. In addition the first two pages of this table give the inverse circular and hyperbolic root-sine transformations. The number of figures provided should be enough for all routine statistical work. Table 10 gives logarithms of factorials to the base 10, useful in calculating individual terms of the binomial and Poisson distributions.

First differences are given wherever they are likely to be helpful in interpolation and a table of proportional parts is provided. In the tables of percentage points instructions on interpolation are given alongside each table.

Tables 3, 5 and 7 are based on *Biometrika* tables and we are indebted to Prof. Pearson and the Editors of *Biometrika* for permission to use them. Additional values not given in the *Biometrika* tables have been specially computed for this collection and our thanks are due to Mr D. A. East and Miss P. A. Johnson who carried out the computations, and to Prof. H. W. Norton for informing us of some calculations of his on the 0·1 % values of F. Table 8 is taken from Kendall and Babington-Smith's collection in the series, *Tracts for Computers*, 24, published by the Department of Statistics, University College, London, and Table 6 from *Tables for Statisticians and Biometricians*, published at the office of *Biometrika* and we are grateful for permission to do this. Finally we should like to thank the staff of the University Press for their helpful advice and co-operation during the printing of the tables.

D. V. LINDLEY
J. C. P. MILLER

9 December 1952

TABLE 1. THE NORMAL DISTRIBUTION FUNCTION

x	$\Phi(x)$		x	$\Phi(x)$		x	$\Phi(x)$		x	$\Phi(x)$		x	$\Phi(x)$	
0·00	0·5000	40	**0·50**	0·6915	35	**1·00**	0·8413	25	**1·50**	0·9332	13	**2·00**	0·97725	53
·01	·5040	40	**·51**	·6950	35	**·01**	·8438	23	**·51**	·9345	12	**·01**	·97778	53
·02	·5080	40	**·52**	·6985	34	**·02**	·8461	24	**·52**	·9357	13	**·02**	·97831	51
·03	·5120	40	**·53**	·7019	35	**·03**	·8485	23	**·53**	·9370	12	**·03**	·97882	50
·04	·5160	39	**·54**	·7054	34	**·04**	·8508	23	**·54**	·9382	12	**·04**	·97932	50
0·05	0·5199	40	**0·55**	0·7088	35	**1·05**	0·8531	23	**1·55**	0·9394	12	**2·05**	0·97982	48
·06	·5239	40	**·56**	·7123	34	**·06**	·8554	23	**·56**	·9406	12	**·06**	·98030	47
·07	·5279	40	**·57**	·7157	33	**·07**	·8577	22	**·57**	·9418	11	**·07**	·98077	47
·08	·5319	40	**·58**	·7190	34	**·08**	·8599	22	**·58**	·9429	12	**·08**	·98124	45
·09	·5359	39	**·59**	·7224	33	**·09**	·8621	22	**·59**	·9441	11	**·09**	·98169	45
0·10	0·5398	40	**0·60**	0·7257	34	**1·10**	0·8643	22	**1·60**	0·9452	11	**2·10**	0·98214	43
·11	·5438	40	**·61**	·7291	33	**·11**	·8665	21	**·61**	·9463	11	**·11**	·98257	43
·12	·5478	39	**·62**	·7324	33	**·12**	·8686	22	**·62**	·9474	10	**·12**	·98300	41
·13	·5517	40	**·63**	·7357	32	**·13**	·8708	21	**·63**	·9484	11	**·13**	·98341	41
·14	·5557	39	**·64**	·7389	33	**·14**	·8729	20	**·64**	·9495	10	**·14**	·98382	40
0·15	0·5596	40	**0·65**	0·7422	32	**1·15**	0·8749	21	**1·65**	0·9505	10	**2·15**	0·98422	39
·16	·5636	39	**·66**	·7454	32	**·16**	·8770	20	**·66**	·9515	10	**·16**	·98461	39
·17	·5675	39	**·67**	·7486	31	**·17**	·8790	20	**·67**	·9525	10	**·17**	·98500	37
·18	·5714	39	**·68**	·7517	32	**·18**	·8810	20	**·68**	·9535	10	**·18**	·98537	37
·19	·5753	40	**·69**	·7549	31	**·19**	·8830	19	**·69**	·9545	9	**·19**	·98574	36
0·20	0·5793	39	**0·70**	0·7580	31	**1·20**	0·8849	20	**1·70**	0·9554	10	**2·20**	0·98610	35
·21	·5832	39	**·71**	·7611	31	**·21**	·8869	19	**·71**	·9564	9	**·21**	·98645	34
·22	·5871	39	**·72**	·7642	31	**·22**	·8888	19	**·72**	·9573	9	**·22**	·98679	34
·23	·5910	38	**·73**	·7673	31	**·23**	·8907	18	**·73**	·9582	9	**·23**	·98713	32
·24	·5948	39	**·74**	·7704	30	**·24**	·8925	19	**·74**	·9591	8	**·24**	·98745	33
0·25	0·5987	39	**0·75**	0·7734	30	**1·25**	0·8944	18	**1·75**	0·9599	9	**2·25**	0·98778	31
·26	·6026	38	**·76**	·7764	30	**·26**	·8962	18	**·76**	·9608	8	**·26**	·98809	31
·27	·6064	39	**·77**	·7794	29	**·27**	·8980	17	**·77**	·9616	9	**·27**	·98840	30
·28	·6103	38	**·78**	·7823	29	**·28**	·8997	18	**·78**	·9625	8	**·28**	·98870	29
·29	·6141	38	**·79**	·7852	29	**·29**	·9015	17	**·79**	·9633	8	**·29**	·98899	29
0·30	0·6179	38	**0·80**	0·7881	29	**1·30**	0·9032	17	**1·80**	0·9641	8	**2·30**	0·98928	28
·31	·6217	38	**·81**	·7910	29	**·31**	·9049	17	**·81**	·9649	7	**·31**	·98956	27
·32	·6255	38	**·82**	·7939	28	**·32**	·9066	16	**·82**	·9656	8	**·32**	·98983	27
·33	·6293	38	**·83**	·7967	28	**·33**	·9082	17	**·83**	·9664	7	**·33**	·99010	26
·34	·6331	37	**·84**	·7995	28	**·34**	·9099	16	**·84**	·9671	7	**·34**	·99036	25
0·35	0·6368	38	**0·85**	0·8023	28	**1·35**	0·9115	16	**1·85**	0·9678	8	**2·35**	0·99061	25
·36	·6406	37	**·86**	·8051	27	**·36**	·9131	16	**·86**	·9686	7	**·36**	·99086	25
·37	·6443	37	**·87**	·8078	28	**·37**	·9147	15	**·87**	·9693	6	**·37**	·99111	23
·38	·6480	37	**·88**	·8106	27	**·38**	·9162	15	**·88**	·9699	7	**·38**	·99134	24
·39	·6517	37	**·89**	·8133	26	**·39**	·9177	15	**·89**	·9706	7	**·39**	·99158	22
0·40	0·6554	37	**0·90**	0·8159	27	**1·40**	0·9192	15	**1·90**	0·9713	6	**2·40**	0·99180	22
·41	·6591	37	**·91**	·8186	26	**·41**	·9207	15	**·91**	·9719	7	**·41**	·99202	22
·42	·6628	36	**·92**	·8212	26	**·42**	·9222	14	**·92**	·9726	6	**·42**	·99224	21
·43	·6664	36	**·93**	·8238	26	**·43**	·9236	15	**·93**	·9732	6	**·43**	·99245	21
·44	·6700	36	**·94**	·8264	25	**·44**	·9251	14	**·94**	·9738	6	**·44**	·99266	20
0·45	0·6736	36	**0·95**	0·8289	26	**1·45**	0·9265	14	**1·95**	0·9744	6	**2·45**	0·99286	19
·46	·6772	36	**·96**	·8315	25	**·46**	·9279	13	**·96**	·9750	6	**·46**	·99305	19
·47	·6808	36	**·97**	·8340	25	**·47**	·9292	14	**·97**	·9756	5	**·47**	·99324	19
·48	·6844	35	**·98**	·8365	24	**·48**	·9306	13	**·98**	·9761	6	**·48**	·99343	18
·49	·6879	36	**·99**	·8389	24	**·49**	·9319	13	**·99**	·9767	5	**·49**	·99361	18
0·50	0·6915		**1·00**	0·8413		**1·50**	0·9332		**2·00**	0·9772		**2·50**	0·99379	

TABLE 1

x	$\Phi(x)$		x	$\Phi(x)$		x	$\Phi(x)$		x	$\phi(x)$	x	$\phi(x)$
2·50	0·99379	17	2·70	0·99653	11	2·90	0·99813	6	0·0	0·3989	2·0	0·0540
·51	·99396	17	·71	·99664	10	·91	·99819	6	0·1	·3970	2·1	·0440
·52	·99413	17	·72	·99674	9	·92	·99825	6	0·2	·3910	2·2	·0355
·53	·99430	16	·73	·99683	10	·93	·99831	5	0·3	·3814	2·3	·0283
·54	·99446	15	·74	·99693	9	·94	·99836	5	0·4	·3683	2·4	·0224
2·55	0·99461	16	2·75	0·99702	9	2·95	0·99841	5	0·5	0·3521	2·5	0·0175
·56	·99477	15	·76	·99711	9	·96	·99846	5	0·6	·3332	2·6	·0136
·57	·99492	14	·77	·99720	8	·97	·99851	5	0·7	·3123	2·7	·0104
·58	·99506	14	·78	·99728	8	·98	·99856	5	0·8	·2897	2·8	·0079
·59	·99520	14	·79	·99736	8	·99	·99861	4	0·9	·2661	2·9	·0060
2·60	0·99534	13	2·80	0·99744	8	3·0	0·99865	38	1·0	0·2420	3·0	0·0044
·61	·99547	13	·81	·99752	8	3·1	·99903	28	1·1	·2179	3·1	·0033
·62	·99560	13	·82	·99760	7	3·2	·99931	21	1·2	·1942	3·2	·0024
·63	·99573	12	·83	·99767	7	3·3	·99952	14	1·3	·1714	3·3	·0017
·64	·99585	13	·84	·99774	7	3·4	·99966	11	1·4	·1497	3·4	·0012
2·65	0·99598	11	2·85	0·99781	7	3·5	0·99977	7	1·5	0·1295	3·5	0·0009
·66	·99609	12	·86	·99788	7	3·6	·99984	5	1·6	·1109	3·6	·0006
·67	·99621	11	·87	·99795	6	3·7	·99989	4	1·7	·0940	3·7	·0004
·68	·99632	11	·88	·99801	6	3·8	·99993	2	1·8	·0790	3·8	·0003
·69	·99643	10	·89	·99807	6	3·9	·99995	2	1·9	·0656	3·9	·0002
2·70	0·99653		2·90	0·99813		4·0	0·99997		2·0	0·0540	4·0	0·0001

The function tabulated is $\Phi(x) = \frac{1}{\sqrt{2\pi}}\int_{-\infty}^{x} e^{-\frac{1}{2}t^2}dt$. $\Phi(x)$ is the probability that a random variable, normally distributed with zero mean and unit variance, will be less than x. The last two columns give the ordinate $\phi(x) = \frac{1}{\sqrt{2\pi}}e^{-\frac{1}{2}x^2}$ of the normal frequency curve.

The critical table below gives on the left the range of values of x for which $\Phi(x)$ takes the value on the right, correct to the last figure given; in critical cases, take the upper of the two values of $\Phi(x)$ indicated.

x	$\Phi(x)$	x	$\Phi(x)$	x	$\Phi(x)$	x	$\Phi(x)$
			0·9994		0·99990		0·99995
3·075		3·263		3·731		3·916	
	0·9990		0·9995		0·99991		0·99996
3·105		3·320		3·759		3·976	
	0·9991		0·9996		0·99992		0·99997
3·138		3·389		3·791		4·055	
	0·9992		0·9997		0·99993		0·99998
3·174		3·480		3·826		4·173	
	0·9993		0·9998		0·99994		0·99999
3·215		3·615		3·867		4·417	
	0·9994		0·9999		0·99995		1·00000

TABLE 2. PERCENTAGE POINTS OF THE NORMAL DISTRIBUTION

P	x	P	x	P	x	P	x	P	x	P	x
50	0·0000	5·0	1·6449	3·0	1·8808	2·0	2·0537	1·0	2·3263	5·0	1·6449
45	0·1257	4·8	1·6646	2·9	1·8957	1·9	2·0749	0·9	2·3656	1·0	2·3263
40	0·2533	4·6	1·6849	2·8	1·9110	1·8	2·0969	0·8	2·4089	0·1	3·0902
35	0·3853	4·4	1·7060	2·7	1·9268	1·7	2·1201	0·7	2·4573	0·01	3·7190
30	0·5244	4·2	1·7279	2·6	1·9431	1·6	2·1444	0·6	2·5121	0·0^{2}1	4·2649
25	0·6745	4·0	1·7507	2·5	1·9600	1·5	2·1701	0·5	2·5758	2·5	1·9600
20	0·8416	3·8	1·7744	2·4	1·9774	1·4	2·1973	0·4	2·6521	0·5	2·5758
15	1·0364	3·6	1·7991	2·3	1·9954	1·3	2·2262	0·3	2·7478	0·05	3·2905
10	1·2816	3·4	1·8250	2·2	2·0141	1·2	2·2571	0·2	2·8782	0·0^{2}5	3·8906
5	1·6449	3·2	1·8522	2·1	2·0335	1·1	2·2904	0·1	3·0902	0·0^{3}5	4·4172

This table gives the percentage points x where $\frac{P}{100} = \frac{1}{\sqrt{2\pi}}\int_{x}^{\infty} e^{-\frac{1}{2}t^2}dt$. The value x is that which is exceeded by a random variable, normally distributed with zero mean and unit variance, with probability $P/100$.

TABLE 3. PERCENTAGE POINTS OF THE t-DISTRIBUTION

P	25	10	5	2	1	0·2	0·1	$\frac{120}{\nu}$
$\nu=1$	2·41	6·31	12·71	31·82	63·66	318·3	636·6	
2	1·60	2·92	4·30	6·96	9·92	22·33	31·60	
3	1·42	2·35	3·18	4·54	5·84	10·21	12·92	
4	1·34	2·13	2·78	3·75	4·60	7·17	8·61	
5	1·30	2·02	2·57	3·36	4·03	5·89	6·87	
6	1·27	1·94	2·45	3·14	3·71	5·21	5·96	
7	1·25	1·89	2·36	3·00	3·50	4·79	5·41	
8	1·24	1·86	2·31	2·90	3·36	4·50	5·04	
9	1·23	1·83	2·26	2·82	3·25	4·30	4·78	
10	1·22	1·81	2·23	2·76	3·17	4·14	4·59	12
12	1·21	1·78	2·18	2·68	3·05	3·93	4·32	10
15	1·20	1·75	2·13	2·60	2·95	3·73	4·07	8
20	1·18	1·72	2·09	2·53	2·85	3·55	3·85	6
24	1·18	1·71	2·06	2·49	2·80	3·47	3·75	5
30	1·17	1·70	2·04	2·46	2·75	3·39	3·65	4
40	1·17	1·68	2·02	2·42	2·70	3·31	3·55	3
60	1·16	1·67	2·00	2·39	2·66	3·23	3·46	2
120	1·16	1·66	1·98	2·36	2·62	3·16	3·37	1
∞	1·15	1·64	1·96	2·33	2·58	3·09	3·29	0

The function tabulated is t_P defined by the equation

$$\frac{P}{100} = \frac{1}{\sqrt{\nu\pi}} \frac{\Gamma(\frac{1}{2}\nu+\frac{1}{2})}{\Gamma(\frac{1}{2}\nu)} \int_{|t|\geqslant t_P} \frac{dt}{(1+t^2/\nu)^{\frac{1}{2}(\nu+1)}}.$$

If t is the ratio of a random variable, normally distributed with zero mean, to an independent estimate of its standard deviation based on ν degrees of freedom, $P/100$ is the probability that $|t| \geqslant t_P$.

Interpolation ν-wise should be linear in $120/\nu$.

Other percentage points may be found approximately, except when ν and P are both small, by using the fact that the variable

$$y = \pm \sinh^{-1}(\sqrt{3t^2/2\nu}),$$

where y has the same sign as t, is approximately normally distributed with zero mean and variance $3/(2\nu-1)$.

TABLE 4. TRANSFORMATION OF THE CORRELATION COEFFICIENT

r	z		r	z		r	z		r	z		r	z		r	z	
0·00	0·000	20	0·40	0·424	24	0·80	1·099	28	0·940	1·738	9	0·960	1·946	13	0·980	2·298	25
·02	·020	20	·42	·448	24	·81	·127	30	·941	·747	9	·961	·959	13	·981	·323	28
·04	·040	20	·44	·472	25	·82	·157	31	·942	·756	8	·962	·972	14	·982	·351	29
·06	·060	20	·46	·497	26	·83	·188	33	·943	·764	10	·963	1·986	14	·983	·380	30
·08	·080	20	·48	·523	26	·84	·221	35	·944	·774	9	·964	2·000	14	·984	·410	33
0·10	0·100	21	0·50	0·549	27	0·85	1·256	37	0·945	1·783	9	0·965	2·014	15	0·985	2·443	34
·12	·121	20	·52	·576	28	·86	·293	40	·946	·792	10	·966	·029	15	·986	·477	38
·14	·141	20	·54	·604	29	·87	·333	43	·947	·802	10	·967	·044	16	·987	·515	40
·16	·161	21	·56	·633	29	·88	·376	46	·948	·812	10	·968	·060	16	·988	·555	44
·18	·182	21	·58	·662	31	·89	·422	50	·949	·822	10	·969	·076	16	·989	·599	48
0·20	0·203	21	0·60	0·693	32	0·90	1·472	56	0·950	1·832	10	0·970	2·092	18	0·990	2·647	53
·22	·224	21	·62	·725	33	·91	·528	61	·951	·842	11	·971	·110	17	·991	·700	59
·24	·245	21	·64	·758	35	·92	·589	69	·952	·853	10	·972	·127	19	·992	·759	67
·26	·266	22	·66	·793	36	·93	·658	80	·953	·863	11	·973	·146	19	·993	·826	77
·28	·288	22	·68	·829	38	·94	·738		·954	·874	12	·974	·165	20	·994	·903	
0·30	0·310	22	0·70	0·867	41	0·95	1·832		0·955	1·886	11	0·975	2·185	20	0·995	2·994	
·32	·332	22	·72	·908	42	·96	1·946		·956	·897	12	·976	·205	22	·996	3·106	
·34	·354	23	·74	·950	46	·97	2·092	see next columns	·957	·909	12	·977	·227	22	·997	·250	
·36	·377	23	·76	0·996	49	·98	·298		·958	·921	12	·978	·249	24	·998	·453	
·38	·400	24	·78	1·045	54	·99	2·647		·959	·933	13	·979	·273	25	·999	3·800	
0·40	0·424		0·80	1·099		1·00	∞		0·960	1·946		0·980	2·298		1·000	∞	

The function tabulated is $z = \tanh^{-1} r = \frac{1}{2}\ln\frac{1+r}{1-r} = 1{\cdot}1513 \log\frac{1+r}{1-r}$.

If r is a partial correlation coefficient, after s variables have been eliminated, in a sample of size n from a multivariate normal population with the corresponding partial correlation coefficient ρ, then z is approximately normally distributed with mean $\tanh^{-1}\rho + \rho/2(n-s-1)$ and variance $1/(n-s-3)$. For $s=0$ we have the ordinary correlation coefficient.

TABLE 5. PERCENTAGE POINTS OF THE χ^2-DISTRIBUTION

P	99·5	99	97·5	95	10	5	2·5	1	0·5	0·1
$\nu=1$	$0{\cdot}0^{4}393$	$0{\cdot}0^{3}157$	$0{\cdot}0^{3}982$	0·00393	2·71	3·84	5·02	6·63	7·88	10·83
2	0·0100	0·0201	0·0506	0·103	4·61	5·99	7·38	9·21	10·60	13·81
3	0·0717	0·115	0·216	0·352	6·25	7·81	9·35	11·34	12·84	16·27
4	0·207	0·297	0·484	0·711	7·78	9·49	11·14	13·28	14·86	18·47
5	0·412	0·554	0·831	1·15	9·24	11·07	12·83	15·09	16·75	20·52
6	0·676	0·872	1·24	1·64	10·64	12·59	14·45	16·81	18·55	22·46
7	0·989	1·24	1·69	2·17	12·02	14·07	16·01	18·48	20·28	24·32
8	1·34	1·65	2·18	2·73	13·36	15·51	17·53	20·09	21·95	26·12
9	1·73	2·09	2·70	3·33	14·68	16·92	19·02	21·67	23·59	27·88
10	2·16	2·56	3·25	3·94	15·99	18·31	20·48	23·21	25·19	29·59
11	2·60	3·05	3·82	4·57	17·28	19·68	21·92	24·73	26·76	31·26
12	3·07	3·57	4·40	5·23	18·55	21·03	23·34	26·22	28·30	32·91
13	3·57	4·11	5·01	5·89	19·81	22·36	24·74	27·69	29·82	34·53
14	4·07	4·66	5·63	6·57	21·06	23·68	26·12	29·14	31·32	36·12
15	4·60	5·23	6·26	7·26	22·31	25·00	27·49	30·58	32·80	37·70
16	5·14	5·81	6·91	7·96	23·54	26·30	28·85	32·00	34·27	39·25
17	5·70	6·41	7·56	8·67	24·77	27·59	30·19	33·41	35·72	40·79
18	6·26	7·01	8·23	9·39	25·99	28·87	31·53	34·81	37·16	42·31
19	6·84	7·63	8·91	10·12	27·20	30·14	32·85	36·19	38·58	43·82
20	7·43	8·26	9·59	10·85	28·41	31·41	34·17	37·57	40·00	45·31
21	8·03	8·90	10·28	11·59	29·62	32·67	35·48	38·93	41·40	46·80
22	8·64	9·54	10·98	12·34	30·81	33·92	36·78	40·29	42·80	48·27
23	9·26	10·20	11·69	13·09	32·01	35·17	38·08	41·64	44·18	49·73
24	9·89	10·86	12·40	13·85	33·20	36·42	39·36	42·98	45·56	51·18
25	10·52	11·52	13·12	14·61	34·38	37·65	40·65	44·31	46·93	52·62
26	11·16	12·20	13·84	15·38	35·56	38·89	41·92	45·64	48·29	54·05
27	11·81	12·88	14·57	16·15	36·74	40·11	43·19	46·96	49·64	55·48
28	12·46	13·56	15·31	16·93	37·92	41·34	44·46	48·28	50·99	56·89
29	13·12	14·26	16·05	17·71	39·09	42·56	45·72	49·59	52·34	58·30
30	13·79	14·95	16·79	18·49	40·26	43·77	46·98	50·89	53·67	59·70
40	20·71	22·16	24·43	26·51	51·81	55·76	59·34	63·69	66·77	73·40
50	27·99	29·71	32·36	34·76	63·17	67·50	71·42	76·15	79·49	86·66
60	35·53	37·48	40·48	43·19	74·40	79·08	83·30	88·38	91·95	99·61
70	43·28	45·44	48·76	51·74	85·53	90·53	95·02	100·4	104·2	112·3
80	51·17	53·54	57·15	60·39	96·58	101·9	106·6	112·3	116·3	124·8
90	59·20	61·75	65·65	69·13	107·6	113·1	118·1	124·1	128·3	137·2
100	67·33	70·06	74·22	77·93	118·5	124·3	129·6	135·8	140·2	149·4

The function tabulated is χ^2_P defined by the equation $\dfrac{P}{100} = \dfrac{1}{2^{\nu/2}\Gamma(\frac{1}{2}\nu)}\displaystyle\int_{\chi^2_P}^{\infty} x^{\frac{1}{2}\nu-1}e^{-x/2}\,dx$. If x is a variable distributed as χ^2 with ν degrees of freedom, $P/100$ is the probability that $x \geqslant \chi^2_P$. For $\nu < 100$, linear interpolation in ν is adequate. For $\nu > 100$, $\sqrt{2\chi^2}$ is approximately normally distributed with mean $\sqrt{2\nu-1}$ and unit variance, and the percentage points may be obtained from Table 2.

TABLE 6. CONVERSION OF RANGE TO STANDARD DEVIATION

n	a_n	n	a_n	n	a_n	n	a_n
2	0·8862	5	0·4299	8	0·3512	11	0·3152
3	0·5908	6	0·3946	9	0·3367	12	0·3069
4	0·4857	7	0·3698	10	0·3249	13	0·2998

An estimate of the standard deviation is given by multiplying the range of a random sample of size n from a normal population, by a_n. The mean range in samples of size n from a normal population is the standard deviation of the population divided by a_n.

TABLE 7 (*a*). 5 PER CENT POINTS OF THE *F*-DISTRIBUTION

$\nu_1=$	1	2	3	4	5	6	7	8	10	12	24	∞
$\nu_2=1$	161·4	199·5	215·7	224·6	230·2	234·0	236·8	238·9	241·9	243·9	249·0	254·3
2	18·5	19·0	19·2	19·2	19·3	19·3	19·4	19·4	19·4	19·4	19·5	19·5
3	10·13	9·55	9·28	9·12	9·01	8·94	8·89	8·85	8·79	8·74	8·64	8·53
4	7·71	6·94	6·59	6·39	6·26	6·16	6·09	6·04	5·96	5·91	5·77	5·63
5	6·61	5·79	5·41	5·19	5·05	4·95	4·88	4·82	4·74	4·68	4·53	4·36
6	5·99	5·14	4·76	4·53	4·39	4·28	4·21	4·15	4·06	4·00	3·84	3·67
7	5·59	4·74	4·35	4·12	3·97	3·87	3·79	3·73	3·64	3·57	3·41	3·23
8	5·32	4·46	4·07	3·84	3·69	3·58	3·50	3·44	3·35	3·28	3·12	2·93
9	5·12	4·26	3·86	3·63	3·48	3·37	3·29	3·23	3·14	3·07	2·90	2·71
10	4·96	4·10	3·71	3·48	3·33	3·22	3·14	3·07	2·98	2·91	2·74	2·54
11	4·84	3·98	3·59	3·36	3·20	3·09	3·01	2·95	2·85	2·79	2·61	2·40
12	4·75	3·89	3·49	3·26	3·11	3·00	2·91	2·85	2·75	2·69	2·51	2·30
13	4·67	3·81	3·41	3·18	3·03	2·92	2·83	2·77	2·67	2·60	2·42	2·21
14	4·60	3·74	3·34	3·11	2·96	2·85	2·76	2·70	2·60	2·53	2·35	2·13
15	4·54	3·68	3·29	3·06	2·90	2·79	2·71	2·64	2·54	2·48	2·29	2·07
16	4·49	3·63	3·24	3·01	2·85	2·74	2·66	2·59	2·49	2·42	2·24	2·01
17	4·45	3·59	3·20	2·96	2·81	2·70	2·61	2·55	2·45	2·38	2·19	1·96
18	4·41	3·55	3·16	2·93	2·77	2·66	2·58	2·51	2·41	2·34	2·15	1·92
19	4·38	3·52	3·13	2·90	2·74	2·63	2·54	2·48	2·38	2·31	2·11	1·88
20	4·35	3·49	3·10	2·87	2·71	2·60	2·51	2·45	2·35	2·28	2·08	1·84
21	4·32	3·47	3·07	2·84	2·68	2·57	2·49	2·42	2·32	2·25	2·05	1·81
22	4·30	3·44	3·05	2·82	2·66	2·55	2·46	2·40	2·30	2·23	2·03	1·78
23	4·28	3·42	3·03	2·80	2·64	2·53	2·44	2·37	2·27	2·20	2·00	1·76
24	4·26	3·40	3·01	2·78	2·62	2·51	2·42	2·36	2·25	2·18	1·98	1·73
25	4·24	3·39	2·99	2·76	2·60	2·49	2·40	2·34	2·24	2·16	1·96	1·71
26	4·23	3·37	2·98	2·74	2·59	2·47	2·39	2·32	2·22	2·15	1·95	1·69
27	4·21	3·35	2·96	2·73	2·57	2·46	2·37	2·31	2·20	2·13	1·93	1·67
28	4·20	3·34	2·95	2·71	2·56	2·45	2·36	2·29	2·19	2·12	1·91	1·65
29	4·18	3·33	2·93	2·70	2·55	2·43	2·35	2·28	2·18	2·10	1·90	1·64
30	4·17	3·32	2·92	2·69	2·53	2·42	2·33	2·27	2·16	2·09	1·89	1·62
32	4·15	3·29	2·90	2·67	2·51	2·40	2·31	2·24	2·14	2·07	1·86	1·59
34	4·13	3·28	2·88	2·65	2·49	2·38	2·29	2·23	2·12	2·05	1·84	1·57
36	4·11	3·26	2·87	2·63	2·48	2·36	2·28	2·21	2·11	2·03	1·82	1·55
38	4·10	3·24	2·85	2·62	2·46	2·35	2·26	2·19	2·09	2·02	1·81	1·53
40	4·08	3·23	2·84	2·61	2·45	2·34	2·25	2·18	2·08	2·00	1·79	1·51
60	4·00	3·15	2·76	2·53	2·37	2·25	2·17	2·10	1·99	1·92	1·70	1·39
120	3·92	3·07	2·68	2·45	2·29	2·18	2·09	2·02	1·91	1·83	1·61	1·25
∞	3·84	3·00	2·60	2·37	2·21	2·10	2·01	1·94	1·83	1·75	1·52	1·00

The function tabulated in Table 7 is F_P defined by the equation

$$\frac{P}{100} = \frac{\Gamma(\frac{1}{2}\nu_1 + \frac{1}{2}\nu_2)}{\Gamma(\frac{1}{2}\nu_1)\,\Gamma(\frac{1}{2}\nu_2)}\,\nu_1^{\frac{1}{2}\nu_1}\,\nu_2^{\frac{1}{2}\nu_2}\int_{F_P}^{\infty}\frac{F^{\frac{1}{2}\nu_1-1}}{(\nu_2+\nu_1 F)^{\frac{1}{2}(\nu_1+\nu_2)}}\,dF,$$

with $P=5$, $2\frac{1}{2}$, 1 and 0·1. If F is the ratio of a mean square on ν_1 degrees of freedom to an independent mean square on ν_2 degrees of freedom, and if the mean squares have equal expectations, then $P/100$ is the probability that $F \geqslant F_P$. The lower percentage points, that is the value F'_P such that $P/100$ is the probability that $F \leqslant F'_P$ may be found by interchanging ν_1 and ν_2 and using the reciprocal of the tabulated value.

Linear interpolation will usually be sufficiently accurate except when either $\nu_1 > 12$ or $\nu_2 > 40$, though occasionally a slight improvement may be effected by using harmonic interpolation. Otherwise, except

TABLE 7 (*b*). $2\frac{1}{2}$ PER CENT POINTS OF THE *F*-DISTRIBUTION

	$\nu_1 =$ 1	2	3	4	5	6	7	8	10	12	24	∞
$\nu_2 =$ 1	648	800	864	900	922	937	948	957	969	977	997	1018
2	38.5	39.0	39.2	39.2	39.3	39.3	39.4	39.4	39.4	39.4	39.5	39.5
3	17.4	16.0	15.4	15.1	14.9	14.7	14.6	14.5	14.4	14.3	14.1	13.9
4	12.22	10.65	9.98	9.60	9.36	9.20	9.07	8.98	8.84	8.75	8.51	8.26
5	10.01	8.43	7.76	7.39	7.15	6.98	6.85	6.76	6.62	6.52	6.28	6.02
6	8.81	7.26	6.60	6.23	5.99	5.82	5.70	5.60	5.46	5.37	5.12	4.85
7	8.07	6.54	5.89	5.52	5.29	5.12	4.99	4.90	4.76	4.67	4.42	4.14
8	7.57	6.06	5.42	5.05	4.82	4.65	4.53	4.43	4.30	4.20	3.95	3.67
9	7.21	5.71	5.08	4.72	4.48	4.32	4.20	4.10	3.96	3.87	3.61	3.33
10	6.94	5.46	4.83	4.47	4.24	4.07	3.95	3.85	3.72	3.62	3.37	3.08
11	6.72	5.26	4.63	4.28	4.04	3.88	3.76	3.66	3.53	3.43	3.17	2.88
12	6.55	5.10	4.47	4.12	3.89	3.73	3.61	3.51	3.37	3.28	3.02	2.72
13	6.41	4.97	4.35	4.00	3.77	3.60	3.48	3.39	3.25	3.15	2.89	2.60
14	6.30	4.86	4.24	3.89	3.66	3.50	3.38	3.29	3.15	3.05	2.79	2.49
15	6.20	4.76	4.15	3.80	3.58	3.41	3.29	3.20	3.06	2.96	2.70	2.40
16	6.12	4.69	4.08	3.73	3.50	3.34	3.22	3.12	2.99	2.89	2.63	2.32
17	6.04	4.62	4.01	3.66	3.44	3.28	3.16	3.06	2.92	2.82	2.56	2.25
18	5.98	4.56	3.95	3.61	3.38	3.22	3.10	3.01	2.87	2.77	2.50	2.19
19	5.92	4.51	3.90	3.56	3.33	3.17	3.05	2.96	2.82	2.72	2.45	2.13
20	5.87	4.46	3.86	3.51	3.29	3.13	3.01	2.91	2.77	2.68	2.41	2.09
21	5.83	4.42	3.82	3.48	3.25	3.09	2.97	2.87	2.73	2.64	2.37	2.04
22	5.79	4.38	3.78	3.44	3.22	3.05	2.93	2.84	2.70	2.60	2.33	2.00
23	5.75	4.35	3.75	3.41	3.18	3.02	2.90	2.81	2.67	2.57	2.30	1.97
24	5.72	4.32	3.72	3.38	3.15	2.99	2.87	2.78	2.64	2.54	2.27	1.94
25	5.69	4.29	3.69	3.35	3.13	2.97	2.85	2.75	2.61	2.51	2.24	1.91
26	5.66	4.27	3.67	3.33	3.10	2.94	2.82	2.73	2.59	2.49	2.22	1.88
27	5.63	4.24	3.65	3.31	3.08	2.92	2.80	2.71	2.57	2.47	2.19	1.85
28	5.61	4.22	3.63	3.29	3.06	2.90	2.78	2.69	2.55	2.45	2.17	1.83
29	5.59	4.20	3.61	3.27	3.04	2.88	2.76	2.67	2.53	2.43	2.15	1.81
30	5.57	4.18	3.59	3.25	3.03	2.87	2.75	2.65	2.51	2.41	2.14	1.79
32	5.53	4.15	3.56	3.22	3.00	2.84	2.72	2.62	2.48	2.38	2.10	1.75
34	5.50	4.12	3.53	3.19	2.97	2.81	2.69	2.59	2.45	2.35	2.08	1.72
36	5.47	4.09	3.51	3.17	2.94	2.79	2.66	2.57	2.43	2.33	2.05	1.69
38	5.45	4.07	3.48	3.15	2.92	2.76	2.64	2.55	2.41	2.31	2.03	1.66
40	5.42	4.05	3.46	3.13	2.90	2.74	2.62	2.53	2.39	2.29	2.01	1.64
60	5.29	3.93	3.34	3.01	2.79	2.63	2.51	2.41	2.27	2.17	1.88	1.48
120	5.15	3.80	3.23	2.89	2.67	2.52	2.39	2.30	2.16	2.05	1.76	1.31
∞	5.02	3.69	3.12	2.79	2.57	2.41	2.29	2.19	2.05	1.94	1.64	1.00

when ν_1 and ν_2 are both large, interpolation should be linear in $\nu_1 F_P$ or $\nu_2 F_P$ (this is equivalent to harmonic interpolation). When ν_1 and ν_2 are both large the percentage points may be found from the formula

$$1.1513 \log F_P = \tfrac{1}{2} \ln F_P = \frac{x_P\sqrt{h+\lambda}}{h} - \left(\frac{1}{\nu_1 - 1} - \frac{1}{\nu_2 - 1}\right)\left(\lambda + \frac{5}{6}\right)$$

where x_P is the P-per cent point of the normal distribution (Table 2), $\lambda = \frac{1}{6}(x_P^2 - 3)$ and $\frac{2}{h} = \frac{1}{\nu_1 - 1} + \frac{1}{\nu_2 - 1}$.

For the values of P given in Table 7, x_P and λ are as follows:

P	5	$2\frac{1}{2}$	1	0.1
x_P	+1.6449	1.9600	2.3263	3.0902
λ	−0.0491	+0.1402	0.4020	1.0916

TABLE 7 (*c*). 1 PER CENT POINTS OF THE *F*-DISTRIBUTION

$\nu_1=$	1	2	3	4	5	6	7	8	10	12	24	∞
$\nu_2=1$	4052	5000	5403	5625	5764	5859	5928	5981	6056	6106	6235	6366
2	98·5	99·0	99·2	99·2	99·3	99·3	99·4	99·4	99·4	99·4	99·5	99·5
3	34·1	30·8	29·5	28·7	28·2	27·9	27·7	27·5	27·2	27·1	26·6	26·1
4	21·2	18·0	16·7	16·0	15·5	15·2	15·0	14·8	14·5	14·4	13·9	13·5
5	16·26	13·27	12·06	11·39	10·97	10·67	10·46	10·29	10·05	9·89	9·47	9·02
6	13·74	10·92	9·78	9·15	8·75	8·47	8·26	8·10	7·87	7·72	7·31	6·88
7	12·25	9·55	8·45	7·85	7·46	7·19	6·99	6·84	6·62	6·47	6·07	5·65
8	11·26	8·65	7·59	7·01	6·63	6·37	6·18	6·03	5·81	5·67	5·28	4·86
9	10·56	8·02	6·99	6·42	6·06	5·80	5·61	5·47	5·26	5·11	4·73	4·31
10	10·04	7·56	6·55	5·99	5·64	5·39	5·20	5·06	4·85	4·71	4·33	3·91
11	9·65	7·21	6·22	5·67	5·32	5·07	4·89	4·74	4·54	4·40	4·02	3·60
12	9·33	6·93	5·95	5·41	5·06	4·82	4·64	4·50	4·30	4·16	3·78	3·36
13	9·07	6·70	5·74	5·21	4·86	4·62	4·44	4·30	4·10	3·96	3·59	3·17
14	8·86	6·51	5·56	5·04	4·70	4·46	4·28	4·14	3·94	3·80	3·43	3·00
15	8·68	6·36	5·42	4·89	4·56	4·32	4·14	4·00	3·80	3·67	3·29	2·87
16	8·53	6·23	5·29	4·77	4·44	4·20	4·03	3·89	3·69	3·55	3·18	2·75
17	8·40	6·11	5·18	4·67	4·34	4·10	3·93	3·79	3·59	3·46	3·08	2·65
18	8·29	6·01	5·09	4·58	4·25	4·01	3·84	3·71	3·51	3·37	3·00	2·57
19	8·18	5·93	5·01	4·50	4·17	3·94	3·77	3·63	3·43	3·30	2·92	2·49
20	8·10	5·85	4·94	4·43	4·10	3·87	3·70	3·56	3·37	3·23	2·86	2·42
21	8·02	5·78	4·87	4·37	4·04	3·81	3·64	3·51	3·31	3·17	2·80	2·36
22	7·95	5·72	4·82	4·31	3·99	3·76	3·59	3·45	3·26	3·12	2·75	2·31
23	7·88	5·66	4·76	4·26	3·94	3·71	3·54	3·41	3·21	3·07	2·70	2·26
24	7·82	5·61	4·72	4·22	3·90	3·67	3·50	3·36	3·17	3·03	2·66	2·21
25	7·77	5·57	4·68	4·18	3·86	3·63	3·46	3·32	3·13	2·99	2·62	2·17
26	7·72	5·53	4·64	4·14	3·82	3·59	3·42	3·29	3·09	2·96	2·58	2·13
27	7·68	5·49	4·60	4·11	3·78	3·56	3·39	3·26	3·06	2·93	2·55	2·10
28	7·64	5·45	4·57	4·07	3·75	3·53	3·36	3·23	3·03	2·90	2·52	2·06
29	7·60	5·42	4·54	4·04	3·73	3·50	3·33	3·20	3·00	2·87	2·49	2·03
30	7·56	5·39	4·51	4·02	3·70	3·47	3·30	3·17	2·98	2·84	2·47	2·01
32	7·50	5·34	4·46	3·97	3·65	3·43	3·26	3·13	2·93	2·80	2·42	1·96
34	7·45	5·29	4·42	3·93	3·61	3·39	3·22	3·09	2·90	2·76	2·38	1·91
36	7·40	5·25	4·38	3·89	3·58	3·35	3·18	3·05	2·86	2·72	2·35	1·87
38	7·35	5·21	4·34	3·86	3·54	3·32	3·15	3·02	2·83	2·69	2·32	1·84
40	7·31	5·18	4·31	3·83	3·51	3·29	3·12	2·99	2·80	2·66	2·29	1·80
60	7·08	4·98	4·13	3·65	3·34	3·12	2·95	2·82	2·63	2·50	2·12	1·60
120	6·85	4·79	3·95	3·48	3·17	2·96	2·79	2·66	2·47	2·34	1·95	1·38
∞	6·63	4·61	3·78	3·32	3·02	2·80	2·64	2·51	2·32	2·18	1·79	1·00

The function tabulated in Table 7 is F_P defined by the equation

$$\frac{P}{100} = \frac{\Gamma(\frac{1}{2}\nu_1+\frac{1}{2}\nu_2)}{\Gamma(\frac{1}{2}\nu_1)\,\Gamma(\frac{1}{2}\nu_2)}\nu_1^{\frac{1}{2}\nu_1}\nu_2^{\frac{1}{2}\nu_2}\int_{F_P}^{\infty}\frac{F^{\frac{1}{2}\nu_1-1}}{(\nu_2+\nu_1F)^{\frac{1}{2}(\nu_1+\nu_2)}}\,dF,$$

with $P=5$, $2\frac{1}{2}$, 1 and 0·1. If F is the ratio of a mean square on ν_1 degrees of freedom to an independent mean square on ν_2 degrees of freedom, and if the mean squares have equal expectations, then $P/100$ is the probability that $F \geqslant F_P$. The lower percentage points, that is the value F_P' such that $P/100$ is the probability that $F \leqslant F_P'$ may be found by interchanging ν_1 and ν_2 and using the reciprocal of the tabulated value.

Linear interpolation will usually be sufficiently accurate except when either $\nu_1 > 12$ or $\nu_2 > 40$, though occasionally a slight improvement may be effected by using harmonic interpolation. Otherwise, except

TABLE 7 (*d*). 0·1 PER CENT POINTS OF THE *F*-DISTRIBUTION

$\nu_1 =$	1	2	3	4	5	6	7	8	10	12	24	∞
$\nu_2 =$ 1*	4053	5000	5404	5625	5764	5859	5929	5981	6056	6107	6235	6366*
2	998·5	999·0	999·2	999·2	999·3	999·3	999·4	999·4	999·4	999·4	999·5	999·5
3	167·0	148·5	141·1	137·1	134·6	132·8	131·5	130·6	129·2	128·3	125·9	123·5
4	74·14	61·25	56·18	53·44	51·71	50·53	49·66	49·00	48·05	47·41	45·77	44·05
5	47·18	37·12	33·20	31·09	29·75	28·83	28·16	27·65	26·92	26·42	25·14	23·79
6	35·51	27·00	23·70	21·92	20·80	20·03	19·46	19·03	18·41	17·99	16·90	15·75
7	29·25	21·69	18·77	17·20	16·21	15·52	15·02	14·63	14·08	13·71	12·73	11·70
8	25·42	18·49	15·83	14·39	13·48	12·86	12·40	12·05	11·54	11·19	10·30	9·34
9	22·86	16·39	13·90	12·56	11·71	11·13	10·69	10·37	9·87	9·57	8·72	7·81
10	21·04	14·91	12·55	11·28	10·48	9·93	9·52	9·20	8·74	8·44	7·64	6·76
11	19·69	13·81	11·56	10·35	9·58	9·05	8·66	8·35	7·92	7·63	6·85	6·00
12	18·64	12·97	10·80	9·63	8·89	8·38	8·00	7·71	7·29	7·00	6·25	5·42
13	17·82	12·31	10·21	9·07	8·35	7·86	7·49	7·21	6·80	6·52	5·78	4·97
14	17·14	11·78	9·73	8·62	7·92	7·44	7·08	6·80	6·40	6·13	5·41	4·60
15	16·59	11·34	9·34	8·25	7·57	7·09	6·74	6·47	6·08	5·81	5·10	4·31
16	16·12	10·97	9·01	7·94	7·27	6·80	6·46	6·19	5·81	5·55	4·85	4·06
17	15·72	10·66	8·73	7·68	7·02	6·56	6·22	5·96	5·58	5·32	4·63	3·85
18	15·38	10·39	8·49	7·46	6·81	6·35	6·02	5·76	5·39	5·13	4·45	3·67
19	15·08	10·16	8·28	7·27	6·62	6·18	5·85	5·59	5·22	4·97	4·29	3·51
20	14·82	9·95	8·10	7·10	6·46	6·02	5·69	5·44	5·08	4·82	4·15	3·38
21	14·59	9·77	7·94	6·95	6·32	5·88	5·56	5·31	4·95	4·70	4·03	3·26
22	14·38	9·61	7·80	6·81	6·19	5·76	5·44	5·19	4·83	4·58	3·92	3·15
23	14·19	9·47	7·67	6·70	6·08	5·65	5·33	5·09	4·73	4·48	3·82	3·05
24	14·03	9·34	7·55	6·59	5·98	5·55	5·23	4·99	4·64	4·39	3·74	2·97
25	13·88	9·22	7·45	6·49	5·89	5·46	5·15	4·91	4·56	4·31	3·66	2·89
26	13·74	9·12	7·36	6·41	5·80	5·38	5·07	4·83	4·48	4·24	3·59	2·82
27	13·61	9·02	7·27	6·33	5·73	5·31	5·00	4·76	4·41	4·17	3·52	2·75
28	13·50	8·93	7·19	6·25	5·66	5·24	4·93	4·69	4·35	4·11	3·46	2·69
29	13·39	8·85	7·12	6·19	5·59	5·18	4·87	4·64	4·29	4·05	3·41	2·64
30	13·29	8·77	7·05	6·12	5·53	5·12	4·82	4·58	4·24	4·00	3·36	2·59
32	13·12	8·64	6·94	6·01	5·43	5·02	4·72	4·48	4·14	3·91	3·27	2·50
34	12·97	8·52	6·83	5·92	5·34	4·93	4·63	4·40	4·06	3·83	3·19	2·42
36	12·83	8·42	6·74	5·84	5·26	4·86	4·56	4·33	3·99	3·76	3·12	2·35
38	12·71	8·33	6·66	5·76	5·19	4·79	4·49	4·26	3·93	3·70	3·06	2·29
40	12·61	8·25	6·59	5·70	5·13	4·73	4·44	4·21	3·87	3·64	3·01	2·23
60	11·97	7·77	6·17	5·31	4·76	4·37	4·09	3·86	3·54	3·32	2·69	1·89
120	11·38	7·32	5·78	4·95	4·42	4·04	3·77	3·55	3·24	3·02	2·40	1·54
∞	10·83	6·91	5·42	4·62	4·10	3·74	3·47	3·27	2·96	2·74	2·13	1·00

when ν_1 and ν_2 are both large, interpolation should be linear in $\nu_1 F_P$ or $\nu_2 F_P$ (this is equivalent to harmonic interpolation). When ν_1 and ν_2 are both large the percentage points may be found from the formula

$$1{\cdot}1513 \log F_P = \tfrac{1}{2} \ln F_P = \frac{x_P\sqrt{h+\lambda}}{h} - \left(\frac{1}{\nu_1 - 1} - \frac{1}{\nu_2 - 1}\right)\left(\lambda + \frac{5}{6}\right)$$

where x_P is the P-per cent point of the normal distribution (Table 2), $\lambda = \frac{1}{6}(x_P^2 - 3)$ and $\frac{2}{h} = \frac{1}{\nu_1 - 1} + \frac{1}{\nu_2 - 1}$. For the values of P given in Table 7, x_P and λ are as follows:

P	5	$2\frac{1}{2}$	1	0·1
x_P	+1·6449	1·9600	2·3263	3·0902
λ	−0·0491	+0·1402	0·4020	1·0916

* Entries for $\nu_2 = 1$ must be multiplied by 100.

TABLE 8. RANDOM SAMPLING NUMBERS

20 17	42 28	23 17	59 66	38 61	02 10	86 10	51 55	92 52	44 25
74 49	04 49	03 04	10 33	53 70	11 54	48 63	94 60	94 49	57 38
94 70	49 31	38 67	23 42	29 65	40 88	78 71	37 18	48 64	06 57
22 15	78 15	69 84	32 52	32 54	15 12	54 02	01 37	38 37	12 93
93 29	12 18	27 30	30 55	91 87	50 57	58 51	49 36	12 53	96 40
45 04	77 97	36 14	99 45	52 95	69 85	03 83	51 87	85 56	22 37
44 91	99 49	89 39	94 60	48 49	06 77	64 72	59 26	08 51	25 57
16 23	91 02	19 96	47 59	89 65	27 84	30 92	63 37	26 24	23 66
04 50	65 04	65 65	82 42	70 51	55 04	61 47	88 83	99 34	82 37
32 70	17 72	03 61	66 26	24 71	22 77	88 33	17 78	08 92	73 49
03 64	59 07	42 95	81 39	06 41	20 81	92 34	51 90	39 08	21 42
62 49	00 90	67 86	93 48	31 83	19 07	67 68	49 03	27 47	52 03
61 00	95 86	98 36	14 03	48 88	51 07	33 40	06 86	33 76	68 57
89 03	90 49	28 74	21 04	09 96	60 45	22 03	52 80	01 79	33 81
01 72	33 85	52 40	60 07	06 71	89 27	14 29	55 24	85 79	31 96
27 56	49 79	34 34	32 22	60 53	91 17	33 26	44 70	93 14	99 70
49 05	74 48	10 55	35 25	24 28	20 22	35 66	66 34	26 35	91 23
49 74	37 25	97 26	33 94	42 23	01 28	59 58	92 69	03 66	73 82
20 26	22 43	88 08	19 85	08 12	47 65	65 63	56 07	97 85	56 79
48 87	77 96	43 39	76 93	08 79	22 18	54 55	93 75	97 26	90 77
08 72	87 46	75 73	00 11	27 07	05 20	30 85	22 21	04 67	19 13
95 97	98 62	17 27	31 42	64 71	46 22	32 75	19 32	20 99	94 85
37 99	57 31	70 40	46 55	46 12	24 32	36 74	69 20	72 10	95 93
05 79	58 37	85 33	75 18	88 71	23 44	54 28	00 48	96 23	66 45
55 85	63 42	00 79	91 22	29 01	41 39	51 40	36 65	26 11	78 32
67 28	96 25	68 36	24 72	03 85	49 24	05 69	64 86	08 19	91 21
85 86	94 78	32 59	51 82	86 43	73 84	45 60	89 57	06 87	08 15
40 10	60 09	05 88	78 44	63 13	58 25	37 11	18 47	75 62	52 21
94 55	89 48	90 80	77 80	26 89	87 44	23 74	66 20	20 19	26 52
11 63	77 77	23 20	33 62	62 19	29 03	94 15	56 37	14 09	47 16
64 00	26 04	54 55	38 57	94 62	68 40	26 04	24 25	03 61	01 20
50 94	13 23	78 41	60 58	10 60	88 46	30 21	45 98	70 96	36 89
66 98	37 96	44 13	45 05	34 59	75 85	48 97	27 19	17 85	48 51
66 91	42 83	60 77	90 91	60 90	79 62	57 66	72 28	08 70	96 03
33 58	12 18	02 07	19 40	21 29	39 45	90 42	58 84	85 43	95 67
52 49	40 16	72 40	73 05	50 90	02 04	98 24	05 30	27 25	20 88
74 98	93 99	78 30	79 47	96 92	45 58	40 37	89 76	84 41	74 68
50 26	54 30	01 88	69 57	54 45	69 88	23 21	05 69	93 44	05 32
49 46	61 89	33 79	96 84	28 34	19 35	28 73	39 59	56 34	97 07
19 65	13 44	78 39	73 88	62 03	36 00	25 96	86 76	67 90	21 68
64 17	47 67	87 59	81 40	72 61	14 00	28 28	55 86	23 38	16 15
18 43	97 37	68 97	56 56	57 95	01 88	11 89	48 07	42 60	11 92
65 58	60 87	51 09	96 61	15 53	66 81	66 88	44 75	37 01	28 88
79 90	31 00	91 14	85 65	31 75	43 15	45 93	64 78	34 53	88 02
07 23	00 15	59 05	16 09	94 42	20 40	63 76	65 67	34 11	94 10
90 08	14 24	01 51	95 46	30 32	33 19	00 14	19 28	40 51	92 69
53 82	62 02	21 82	34 13	41 03	12 85	65 30	00 97	56 30	15 48
98 17	26 15	04 50	76 25	20 33	54 84	39 31	23 33	59 64	96 27
08 91	12 44	82 40	30 62	45 50	64 54	65 17	89 25	59 44	99 95
37 21	46 77	84 87	67 39	85 54	97 37	33 41	11 74	90 50	29 62

Each digit is an independent sample from a population in which the digits 0 to 9 are equally likely, that is each has a probability of $\frac{1}{10}$.

TABLE 8

16 16	57 04	81 71	17 46	53 29	73 46	42 73	77 63	62 58	60 59
98 63	89 52	77 23	61 08	63 90	80 38	42 71	85 70	04 81	05 50
01 03	09 35	02 54	51 96	92 75	58 29	24 23	25 19	89 97	91 29
29 07	16 34	49 22	52 96	89 34	17 11	06 91	24 38	55 06	83 59
72 61	80 54	70 99	24 64	11 38	83 65	27 23	40 37	84 58	48 53
71 11	41 82	79 37	00 45	98 54	52 89	26 34	40 13	60 38	08 86
61 05	66 18	76 82	11 18	61 90	90 63	78 57	32 06	39 95	75 94
81 89	42 34	00 49	97 53	33 16	26 91	57 58	42 48	51 05	48 27
10 24	90 84	22 16	26 96	54 11	01 96	58 81	37 97	80 98	72 81
14 28	33 43	01 32	58 39	19 54	56 57	23 58	24 87	77 36	20 97
35 41	17 89	87 04	28 32	13 45	59 03	91 08	69 24	84 44	42 83
07 89	36 87	98 73	77 64	75 19	05 61	11 64	31 75	49 38	96 60
27 59	15 58	19 68	95 47	25 69	11 90	26 19	07 40	83 59	90 95
95 98	45 52	27 35	86 81	16 29	37 60	39 35	05 24	49 00	29 07
12 95	72 72	81 84	36 58	05 10	70 50	31 04	12 67	74 01	72 90
35 23	06 68	52 50	39 55	92 28	28 89	64 87	80 00	84 53	97 97
86 33	95 73	80 92	26 49	54 50	41 21	06 62	73 91	35 05	21 37
02 82	96 23	16 46	15 51	60 31	55 27	84 14	71 58	94 71	48 35
44 46	34 96	32 68	48 22	40 17	43 25	33 31	26 26	59 34	99 00
08 77	07 19	94 46	17 51	03 73	99 89	28 44	16 87	56 16	56 09
61 59	37 08	08 46	56 76	29 48	33 87	70 79	03 80	96 81	79 68
67 70	18 01	67 19	29 49	58 67	08 56	27 24	20 70	46 31	04 32
23 09	08 79	18 78	00 32	86 74	78 55	55 72	58 54	76 07	53 73
89 40	26 39	74 58	59 55	87 11	74 06	49 46	31 94	86 66	66 97
84 95	66 42	90 74	13 71	00 71	24 41	67 62	38 92	39 26	30 29
52 14	49 02	19 31	28 15	51 01	19 09	97 94	52 43	22 21	17 66
89 56	31 41	37 87	28 16	62 48	01 84	46 06	04 39	94 10	76 21
65 94	05 93	06 68	34 72	73 17	65 34	00 65	75 78	23 97	13 04
13 08	15 75	02 83	48 26	53 77	62 96	56 52	28 26	12 15	75 53
03 18	33 57	16 71	60 27	15 18	39 32	37 01	05 86	25 14	35 41
10 04	00 95	85 04	32 80	19 01	85 03	29 29	80 04	21 52	14 76
23 94	97 28	60 43	42 25	26 48	48 13	34 68	39 22	74 85	03 25
35 63	42 90	90 74	33 17	58 77	83 36	76 22	00 89	61 55	13 17
42 86	03 36	45 33	60 77	72 92	10 76	22 55	11 00	37 60	47 73
67 26	92 87	09 96	85 37	82 61	39 01	70 05	12 66	17 39	99 34
91 93	88 56	35 76	97 35	19 37	14 66	07 57	24 41	06 90	07 72
37 14	73 35	32 01	07 94	78 28	90 33	71 56	63 77	89 24	24 28
07 46	50 58	08 73	42 97	20 42	64 68	48 35	04 38	28 28	36 94
92 18	09 46	94 99	17 41	28 60	67 94	26 54	63 70	84 73	76 61
00 49	98 43	39 67	68 40	41 31	92 28	49 57	15 55	11 81	41 89
08 59	41 41	33 59	43 28	14 51	02 71	24 45	41 57	22 11	79 79
67 05	19 54	32 33	34 68	27 93	39 35	62 51	35 55	40 99	46 19
24 99	48 06	96 41	21 25	29 03	57 71	96 49	94 74	98 90	21 52
65 86	27 46	70 93	27 39	64 37	01 63	21 03	43 78	18 74	77 07
52 70	03 20	84 96	14 37	51 05	63 99	81 02	84 56	17 78	48 45
32 88	29 93	58 21	71 05	68 58	79 08	86 37	98 76	70 45	66 23
54 16	39 40	98 57	02 05	65 15	73 23	51 51	75 06	38 13	51 68
95 22	18 59	54 57	44 22	72 35	81 24	14 94	24 04	42 26	92 14
93 10	27 94	90 45	39 33	50 26	88 46	90 57	40 47	71 63	62 59
19 20	85 20	15 67	78 03	32 23	50 59	24 83	64 99	18 00	78 50

Each digit is an independent sample from a population in which the digits 0 to 9 are equally likely, that is each has a probability of $\frac{1}{10}$.

TABLE 9

n	n^2	$\sqrt{n}$	$\frac{1}{n}$	$\frac{1}{\sqrt{n}}$	x	$\sin^{-1}\sqrt{x}$		$\sinh^{-1}$ $\sqrt{x}$		$\sinh^{-1}$ $\sqrt{10x}$		$\sinh^{-1}$ $\sqrt{100x}$		$\log t$	t	
0	0	0·0000			0·00	0·000	*	0·000	*	0·000	see previous column	0·000	see previous column	0·000	1·0000	23
1	1	1·0000	1·00000	1·00000	·01	·100	*	·100	*	·311		0·881		·001	·0023	23
2	4	1·4142	0·50000	0·70711	·02	·142	32	·141	31	·434		1·146		·002	·0046	23
3	9	1·7321	·33333	·57735	·03	·174	27	·172	27	·523		·317		·003	·0069	24
4	16	2·0000	·25000	·50000	·04	·201	25	·199	23	·596		·444		·004	·0093	23
5	25	2·2361	0·20000	0·44721	0·05	0·226	21	0·222	21	0·658		1·544		0·005	1·0116	23
6	36	2·4495	·16667	·40825	·06	·247	21	·243	19	·713		·628		·006	·0139	23
7	49	2·6458	·14286	·37796	·07	·268	19	·262	17	·761		·700		·007	·0162	24
8	64	2·8284	·12500	·35355	·08	·287	18	·279	17	·805		·763		·008	·0186	23
9	81	3·0000	·11111	·33333	·09	·305	17	·296	15	·845		·818		·009	·0209	24
10	1 00	3·1623	0·10000	0·31623	0·10	0·322	16	0·311	15	0·881	34	1·869	45	0·010	1·0233	24
11	1 21	3·3166	·09091	·30151	·11	·338	16	·326	14	·915	32	·914	42	·011	·0257	23
12	1 44	3·4641	·08333	·28868	·12	·354	15	·340	13	·947	30	·956	38	·012	·0280	24
13	1 69	3·6056	·07692	·27735	·13	·369	14	·353	13	0·977	28	1·994	36	·013	·0304	24
14	1 96	3·7417	·07143	·26726	·14	·383	15	·366	12	1·005	27	2·030	33	·014	·0328	23
15	2 25	3·8730	0·06667	0·25820	0·15	0·398	14	0·378	12	1·032	25	2·063	32	0·015	1·0351	24
16	2 56	4·0000	·06250	·25000	·16	·412	13	·390	11	·057	24	·095	29	·016	·0375	24
17	2 89	4·1231	·05882	·24254	·17	·425	13	·401	11	·081	23	·124	28	·017	·0399	24
18	3 24	4·2426	·05556	·23570	·18	·438	13	·412	11	·104	21	·152	26	·018	·0423	24
19	3 61	4·3589	·05263	·22942	·19	·451	13	·423	11	·125	21	·178	25	·019	·0447	24
20	4 00	4·4721	0·05000	0·22361	0·20	0·464	12	0·434	10	1·146	20	2·203	24	0·020	1·0471	24
21	4 41	4·5826	·04762	·21822	·21	·476	12	·444	9	·166	19	·227	23	·021	·0495	25
22	4 84	4·6904	·04545	·21320	·22	·488	12	·453	10	·185	19	·250	22	·022	·0520	24
23	5 29	4·7958	·04348	·20851	·23	·500	12	·463	9	·204	18	·272	20	·023	·0544	24
24	5 76	4·8990	·04167	·20412	·24	·512	12	·472	9	·222	17	·292	20	·024	·0568	25
25	6 25	5·0000	0·04000	0·20000	0·25	0·524	11	0·481	9	1·239	17	2·312	20	0·025	1·0593	24
26	6 76	5·0990	·03846	·19612	·26	·535	11	·490	9	·256	16	·332	18	·026	·0617	24
27	7 29	5·1962	·03704	·19245	·27	·546	12	·499	8	·272	15	·350	18	·027	·0641	25
28	7 84	5·2915	·03571	·18898	·28	·558	11	·507	8	·287	15	·368	17	·028	·0666	25
29	8 41	5·3852	·03448	·18570	·29	·569	11	·515	8	·302	15	·385	17	·029	·0691	24
30	9 00	5·4772	0·03333	0·18257	0·30	0·580	11	0·523	8	1·317	14	2·402	16	0·030	1·0715	25
31	9 61	5·5678	·03226	·17961	·31	·591	10	·531	8	·331	14	·418	16	·031	·0740	25
32	10 24	5·6569	·03125	·17678	·32	·601	11	·539	8	·345	13	·434	15	·032	·0765	24
33	10 89	5·7446	·03030	·17408	·33	·612	11	·547	7	·358	14	·449	15	·033	·0789	25
34	11 56	5·8310	·02941	·17150	·34	·623	10	·554	8	·372	12	·464	14	·034	·0814	25
35	12 25	5·9161	0·02857	0·16903	0·35	0·633	11	0·562	7	1·384	13	2·478	14	0·035	1·0839	25
36	12 96	6·0000	·02778	·16667	·36	·644	10	·569	7	·397	12	·492	13	·036	·0864	25
37	13 69	6·0828	·02703	·16440	·37	·654	10	·576	7	·409	12	·505	13	·037	·0889	25
38	14 44	6·1644	·02632	·16222	·38	·664	10	·583	7	·421	11	·518	13	·038	·0914	26
39	15 21	6·2450	·02564	·16013	·39	·674	11	·590	6	·432	12	·531	13	·039	·0940	25
40	16 00	6·3246	0·02500	0·15811	0·40	0·685	10	0·596	7	1·444	11	2·544	12	0·040	1·0965	25
41	16 81	6·4031	·02439	·15617	·41	·695	10	·603	7	·455	11	·556	12	·041	·0990	25
42	17 64	6·4807	·02381	·15430	·42	·705	10	·610	6	·466	10	·568	12	·042	·1015	26
43	18 49	6·5574	·02326	·15250	·43	·715	10	·616	6	·476	10	·580	11	·043	·1041	25
44	19 36	6·6332	·02273	·15076	·44	·725	10	·622	7	·486	11	·591	11	·044	·1066	26
45	20 25	6·7082	0·02222	0·14907	0·45	0·735	10	0·629	6	1·497	10	2·602	11	0·045	1·1092	25
46	21 16	6·7823	·02174	·14744	·46	·745	10	·635	6	·507	9	·613	10	·046	·1117	26
47	22 09	6·8557	·02128	·14586	·47	·755	10	·641	6	·516	10	·623	11	·047	·1143	26
48	23 04	6·9282	·02083	·14434	·48	·765	10	·647	6	·526	9	·634	10	·048	·1169	25
49	24 01	7·0000	·02041	·14286	·49	·775	10	·653	5	·535	9	·644	10	·049	·1194	26
50	25 00	7·0711	0·02000	0·14142	0·50	0·785		0·658		1·544		2·654		0·050	1·1220	

* For small x $\sin^{-1}\sqrt{x} \doteqdot \sqrt{x}$ $\sinh^{-1}\sqrt{x} \doteqdot \sqrt{x}$

TABLE 9

n	n^2	$\sqrt{n}$	$\frac{1}{n}$	$\frac{1}{\sqrt{n}}$	x	$\sin^{-1}\sqrt{x}$	Δ	$\sinh^{-1}$ $\sqrt{x}$	Δ	$\sinh^{-1}$ $\sqrt{10x}$	Δ	$\sinh^{-1}$ $\sqrt{100x}$	Δ	$\log t$	t	Δ
50	25 00	7·0711	0·02000	0·14142	0·50	0·785	10	0·658	6	1·544	10	2·654	10	0·050	1·1220	26
51	26 01	7·1414	·01961	·14003	·51	·795	10	·664	6	·554	8	·664	10	·051	·1246	26
52	27 04	7·2111	·01923	·13868	·52	·805	10	·670	5	·562	9	·674	9	·052	·1272	26
53	28 09	7·2801	·01887	·13736	·53	·815	10	·675	6	·571	9	·683	9	·053	·1298	26
54	29 16	7·3485	·01852	·13608	·54	·825	10	·681	5	·580	8	·692	9	·054	·1324	26
55	30 25	7·4162	0·01818	0·13484	0·55	0·835	11	0·686	6	1·588	8	2·701	9	0·055	1·1350	26
56	31 36	7·4833	·01786	·13363	·56	·846	10	·692	5	·596	9	·710	9	·056	·1376	26
57	32 49	7·5498	·01754	·13245	·57	·856	10	·697	5	·605	8	·719	9	·057	·1402	27
58	33 64	7·6158	·01724	·13131	·58	·866	10	·702	6	·613	8	·728	8	·058	·1429	26
59	34 81	7·6811	·01695	·13019	·59	·876	10	·708	5	·621	7	·736	8	·059	·1455	27
60	36 00	7·7460	0·01667	0·12910	0·60	0·886	10	0·713	5	1·628	8	2·744	9	0·060	1·1482	26
61	37 21	7·8102	·01639	·12804	·61	·896	11	·718	5	·636	8	·753	8	·061	·1508	27
62	38 44	7·8740	·01613	·12700	·62	·907	10	·723	5	·644	7	·761	8	·062	·1535	26
63	39 69	7·9373	·01587	·12599	·63	·917	10	·728	5	·651	7	·769	7	·063	·1561	27
64	40 96	8·0000	·01562	·12500	·64	·927	11	·733	5	·658	7	·776	8	·064	·1588	26
65	42 25	8·0623	0·01538	0·12403	0·65	0·938	10	0·738	4	1·665	8	2·784	8	0·065	1·1614	27
66	43 56	8·1240	·01515	·12309	·66	·948	11	·742	5	·673	7	·792	7	·066	·1641	27
67	44 89	8·1854	·01493	·12217	·67	·959	11	·747	5	·680	6	·799	8	·067	·1668	27
68	46 24	8·2462	·01471	·12127	·68	·970	10	·752	4	·686	7	·807	7	·068	·1695	27
69	47 61	8·3066	·01449	·12039	·69	·980	11	·756	5	·693	7	·814	7	·069	·1722	27
70	49 00	8·3666	0·01429	0·11952	0·70	0·991	11	0·761	5	1·700	7	2·821	7	0·070	1·1749	27
71	50 41	8·4261	·01408	·11868	·71	1·002	11	·766	4	·707	6	·828	7	·071	·1776	27
72	51 84	8·4853	·01389	·11785	·72	·013	11	·770	5	·713	7	·835	7	·072	·1803	27
73	53 29	8·5440	·01370	·11704	·73	·024	12	·775	4	·720	6	·842	7	·073	·1830	28
74	54 76	8·6023	·01351	·11625	·74	·036	11	·779	4	·726	6	·849	6	·074	·1858	27
75	56 25	8·6603	0·01333	0·11547	0·75	1·047	12	0·783	5	1·732	7	2·855	7	0·075	1·1885	27
76	57 76	8·7178	·01316	·11471	·76	·059	12	·788	4	·739	6	·862	6	·076	·1912	28
77	59 29	8·7750	·01299	·11396	·77	·071	12	·792	4	·745	6	·868	7	·077	·1940	27
78	60 84	8·8318	·01282	·11323	·78	·083	12	·796	5	·751	6	·875	6	·078	·1967	28
79	62 41	8·8882	·01266	·11251	·79	·095	12	·801	4	·757	6	·881	6	·079	·1995	28
80	64 00	8·9443	0·01250	0·11180	0·80	1·107	13	0·805	4	1·763	6	2·887	6	0·080	1·2023	27
81	65 61	9·0000	·01235	·11111	·81	·120	13	·809	4	·769	5	·893	7	·081	·2050	28
82	67 24	9·0554	·01220	·11043	·82	·133	13	·813	4	·774	6	·900	6	·082	·2078	28
83	68 89	9·1104	·01205	·10976	·83	·146	13	·817	4	·780	6	·906	6	·083	·2106	28
84	70 56	9·1652	·01190	·10911	·84	·159	14	·821	4	·786	5	·912	5	·084	·2134	28
85	72 25	9·2195	0·01176	0·10847	0·85	1·173	14	0·825	4	1·791	6	2·917	6	0·085	1·2162	28
86	73 96	9·2736	·01163	·10783	·86	·187	15	·829	4	·797	5	·923	6	·086	·2190	28
87	75 69	9·3274	·01149	·10721	·87	·202	15	·833	4	·802	6	·929	6	·087	·2218	28
88	77 44	9·3808	·01136	·10660	·88	·217	16	·837	4	·808	5	·935	5	·088	·2246	28
89	79 21	9·4340	·01124	·10600	·89	·233	16	·841	4	·813	5	·940	6	·089	·2274	29
90	81 00	9·4868	0·01111	0·10541	0·90	1·249	17	0·845	3	1·818	6	2·946	5	0·090	1·2303	28
91	82 81	9·5394	·01099	·10483	·91	·266	18	·848	4	·824	5	·951	6	·091	·2331	28
92	84 64	9·5917	·01087	·10426	·92	·284	19	·852	4	·829	5	·957	5	·092	·2359	29
93	86 49	9·6437	·01075	·10370	·93	·303	20	·856	4	·834	5	·962	5	·093	·2388	29
94	88 36	9·6954	·01064	·10314	·94	·323	22	·860	3	·839	5	·967	6	·094	·2417	28
95	90 25	9·7468	0·01053	0·10260	0·95	1·345	24	0·863	4	1·844	5	2·973	5	0·095	1·2445	29
96	92 16	9·7980	·01042	·10206	·96	·369	28	·867	4	·849	5	·978	5	·096	·2474	29
97	94 09	9·8489	·01031	·10153	·97	·397	32	·871	3	·854	5	·983	5	·097	·2503	28
98	96 04	9·8995	·01020	·10102	·98	·429	•	·874	4	·859	5	·988	5	·098	·2531	29
99	98 01	9·9499	·01010	·10050	·99	·471	•	·878	3	·864	5	·993	5	·099	·2560	29
100	100 00	10·0000	0·01000	0·10000	1·00	1·571		0·881		1·869		2·998		0·100	1·2589	

• For x near unity $\sin^{-1}\sqrt{x} \doteqdot 1{\cdot}571 - \sqrt{1-x}$

TABLE 9

x	x^2	$\sqrt{x}$	Δ	$\sqrt{10x}$	Δ	$\frac{1}{x}$	Δ	$\frac{1}{\sqrt{x}}$	Δ	$\frac{1}{\sqrt{10x}}$	Δ	$\log x$	Δ	$\log t$	t	Δ
1·00	1·0000	1·0000	50	3·162	16	1·0000	99	1·0000	50	0·3162	15	0·0000	43	**0·100**	1·2589	29
·01	1·0201	·0050	50	·178	16	0·9901	97	0·9950	49	·3147	16	·0043	43	**·101**	·2618	29
·02	1·0404	·0100	49	·194	15	·9804	95	·9901	48	·3131	15	·0086	42	**·102**	·2647	30
·03	1·0609	·0149	49	·209	16	·9709	94	·9853	47	·3116	15	·0128	42	**·103**	·2677	29
·04	1·0816	·0198	49	·225	15	·9615	91	·9806	47	·3101	15	·0170	42	**·104**	·2706	29
1·05	1·1025	1·0247	49	3·240	16	0·9524	90	0·9759	46	0·3086	15	0·0212	41	**0·105**	1·2735	29
·06	1·1236	·0296	48	·256	15	·9434	88	·9713	46	·3071	14	·0253	41	**·106**	·2764	30
·07	1·1449	·0344	48	·271	15	·9346	87	·9667	44	·3057	14	·0294	40	**·107**	·2794	29
·08	1·1664	·0392	48	·286	16	·9259	85	·9623	45	·3043	14	·0334	40	**·108**	·2823	30
·09	1·1881	·0440	48	·302	15	·9174	83	·9578	43	·3029	14	·0374	40	**·109**	·2853	29
1·10	1·2100	1·0488	48	3·317	15	0·9091	82	0·9535	43	0·3015	13	0·0414	39	**0·110**	1·2882	30
·11	1·2321	·0536	47	·332	15	·9009	80	·9492	43	·3002	14	·0453	39	**·111**	·2912	30
·12	1·2544	·0583	47	·347	15	·8929	79	·9449	42	·2988	13	·0492	39	**·112**	·2942	30
·13	1·2769	·0630	47	·362	14	·8850	78	·9407	41	·2975	13	·0531	38	**·113**	·2972	30
·14	1·2996	·0677	47	·376	15	·8772	76	·9366	41	·2962	13	·0569	38	**·114**	·3002	30
1·15	1·3225	1·0724	46	3·391	15	0·8696	75	0·9325	40	0·2949	13	0·0607	38	**0·115**	1·3032	30
·16	1·3456	·0770	47	·406	15	·8621	74	·9285	40	·2936	12	·0645	37	**·116**	·3062	30
·17	1·3689	·0817	46	·421	14	·8547	72	·9245	39	·2924	13	·0682	37	**·117**	·3092	30
·18	1·3924	·0863	46	·435	15	·8475	72	·9206	39	·2911	12	·0719	36	**·118**	·3122	30
·19	1·4161	·0909	45	·450	14	·8403	70	·9167	38	·2899	12	·0755	37	**·119**	·3152	31
1·20	1·4400	1·0954	46	3·464	15	0·8333	69	0·9129	38	0·2887	12	0·0792	36	**0·120**	1·3183	30
·21	1·4641	·1000	45	·479	14	·8264	67	·9091	37	·2875	12	·0828	36	**·121**	·3213	30
·22	1·4884	·1045	46	·493	14	·8197	67	·9054	37	·2863	12	·0864	35	**·122**	·3243	31
·23	1·5129	·1091	45	·507	14	·8130	65	·9017	37	·2851	11	·0899	35	**·123**	·3274	31
·24	1·5376	·1136	44	·521	15	·8065	65	·8980	36	·2840	12	·0934	35	**·124**	·3305	30
1·25	1·5625	1·1180	45	3·536	14	0·8000	63	0·8944	35	0·2828	11	0·0969	35	**0·125**	1·3335	31
·26	1·5876	·1225	44	·550	14	·7937	63	·8909	35	·2817	11	·1004	34	**·126**	·3366	31
·27	1·6129	·1269	45	·564	14	·7874	62	·8874	35	·2806	11	·1038	34	**·127**	·3397	31
·28	1·6384	·1314	44	·578	14	·7812	60	·8839	34	·2795	11	·1072	34	**·128**	·3428	31
·29	1·6641	·1358	44	·592	14	·7752	60	·8805	34	·2784	10	·1106	33	**·129**	·3459	31
1·30	1·6900	1·1402	44	3·606	13	0·7692	58	0·8771	34	0·2774	11	0·1139	34	**0·130**	1·3490	31
·31	1·7161	·1446	43	·619	14	·7634	58	·8737	33	·2763	11	·1173	33	**·131**	·3521	31
·32	1·7424	·1489	44	·633	14	·7576	57	·8704	33	·2752	10	·1206	33	**·132**	·3552	31
·33	1·7689	·1533	43	·647	14	·7519	56	·8671	32	·2742	10	·1239	32	**·133**	·3583	31
·34	1·7956	·1576	43	·661	13	·7463	56	·8639	32	·2732	10	·1271	32	**·134**	·3614	32
1·35	1·8225	1·1619	43	3·674	14	0·7407	54	0·8607	32	0·2722	10	0·1303	32	**0·135**	1·3646	31
·36	1·8496	·1662	43	·688	13	·7353	54	·8575	31	·2712	10	·1335	32	**·136**	·3677	32
·37	1·8769	·1705	42	·701	14	·7299	53	·8544	31	·2702	10	·1367	32	**·137**	·3709	31
·38	1·9044	·1747	43	·715	13	·7246	52	·8513	31	·2692	10	·1399	31	**·138**	·3740	32
·39	1·9321	·1790	42	·728	14	·7194	51	·8482	30	·2682	9	·1430	31	**·139**	·3772	32
1·40	1·9600	1·1832	42	3·742	13	0·7143	51	0·8452	30	0·2673	10	0·1461	31	**0·140**	1·3804	32
·41	1·9881	·1874	42	·755	13	·7092	50	·8422	30	·2663	9	·1492	31	**·141**	·3836	32
·42	2·0164	·1916	42	·768	14	·7042	49	·8392	30	·2654	10	·1523	30	**·142**	·3868	32
·43	2·0449	·1958	42	·782	13	·6993	49	·8362	29	·2644	9	·1553	31	**·143**	·3900	32
·44	2·0736	·2000	42	·795	13	·6944	47	·8333	28	·2635	9	·1584	30	**·144**	·3932	32
1·45	2·1025	1·2042	41	3·808	13	0·6897	48	0·8305	29	0·2626	9	0·1614	30	**0·145**	1·3964	32
·46	2·1316	·2083	41	·821	13	·6849	46	·8276	28	·2617	9	·1644	29	**·146**	·3996	32
·47	2·1609	·2124	42	·834	13	·6803	46	·8248	28	·2608	9	·1673	30	**·147**	·4028	32
·48	2·1904	·2166	41	·847	13	·6757	46	·8220	28	·2599	8	·1703	29	**·148**	·4060	33
·49	2·2201	·2207	40	·860	13	·6711	44	·8192	27	·2591	9	·1732	29	**·149**	·4093	32
1·50	2·2500	1·2247		3·873		0·6667		0·8165		0·2582		0·1761		**0·150**	1·4125	

TABLE 9

x	x^2	$\sqrt{x}$		$\sqrt{10x}$		$\frac{1}{x}$		$\frac{1}{\sqrt{x}}$		$\frac{1}{\sqrt{10x}}$		$\log x$		$\log t$	t	
1·50	2·2500	1·2247	41	3·873	13	0·6667	44	0·8165	27	0·2582	9	0·1761	29	0·150	1·4125	33
·51	2·2801	·2288	41	·886	13	·6623	44	·8138	27	·2573	8	·1790	28	·151	·4158	33
·52	2·3104	·2329	40	·899	13	·6579	43	·8111	26	·2565	8	·1818	29	·152	·4191	32
·53	2·3409	·2369	41	·912	12	·6536	42	·8085	27	·2557	9	·1847	28	·153	·4223	33
·54	2·3716	·2410	40	·924	13	·6494	42	·8058	26	·2548	8	·1875	28	·154	·4256	33
1·55	2·4025	1·2450	40	3·937	13	0·6452	42	0·8032	26	0·2540	8	0·1903	28	0·155	1·4289	33
·56	2·4336	·2490	40	·950	12	·6410	41	·8006	25	·2532	8	·1931	28	·156	·4322	33
·57	2·4649	·2530	40	·962	13	·6369	40	·7981	25	·2524	8	·1959	28	·157	·4355	33
·58	2·4964	·2570	40	·975	12	·6329	40	·7956	25	·2516	8	·1987	27	·158	·4388	33
·59	2·5281	·2610	39	3·987	13	·6289	39	·7931	25	·2508	8	·2014	27	·159	·4421	33
1·60	2·5600	1·2649	40	4·000	12	0·6250	39	0·7906	25	0·2500	8	0·2041	27	0·160	1·4454	34
·61	2·5921	·2689	39	·012	13	·6211	38	·7881	24	·2492	7	·2068	27	·161	·4488	33
·62	2·6244	·2728	39	·025	12	·6173	38	·7857	24	·2485	8	·2095	27	·162	·4521	34
·63	2·6569	·2767	39	·037	13	·6135	37	·7833	24	·2477	8	·2122	26	·163	·4555	33
·64	2·6896	·2806	39	·050	12	·6098	37	·7809	24	·2469	7	·2148	27	·164	·4588	34
1·65	2·7225	1·2845	39	4·062	12	0·6061	37	0·7785	23	0·2462	8	0·2175	26	0·165	1·4622	33
·66	2·7556	·2884	39	·074	13	·6024	36	·7762	24	·2454	7	·2201	26	·166	·4655	34
·67	2·7889	·2923	38	·087	12	·5988	36	·7738	23	·2447	7	·2227	26	·167	·4689	34
·68	2·8224	·2961	39	·099	12	·5952	35	·7715	23	·2440	7	·2253	26	·168	·4723	34
·69	2·8561	·3000	38	·111	12	·5917	35	·7692	22	·2433	8	·2279	25	·169	·4757	34
1·70	2·8900	1·3038	39	4·123	12	0·5882	34	0·7670	23	0·2425	7	0·2304	26	0·170	1·4791	34
·71	2·9241	·3077	38	·135	12	·5848	34	·7647	22	·2418	7	·2330	25	·171	·4825	34
·72	2·9584	·3115	38	·147	12	·5814	34	·7625	22	·2411	7	·2355	25	·172	·4859	35
·73	2·9929	·3153	38	·159	12	·5780	33	·7603	22	·2404	7	·2380	25	·173	·4894	34
·74	3·0276	·3191	38	·171	12	·5747	33	·7581	22	·2397	7	·2405	25	·174	·4928	34
1·75	3·0625	1·3229	37	4·183	12	0·5714	32	0·7559	21	0·2390	6	0·2430	25	0·175	1·4962	35
·76	3·0976	·3266	38	·195	12	·5682	32	·7538	22	·2384	7	·2455	25	·176	·4997	34
·77	3·1329	·3304	38	·207	12	·5650	32	·7516	21	·2377	7	·2480	24	·177	·5031	35
·78	3·1684	·3342	37	·219	12	·5618	31	·7495	21	·2370	6	·2504	25	·178	·5066	35
·79	3·2041	·3379	37	·231	12	·5587	31	·7474	20	·2364	7	·2529	24	·179	·5101	35
1·80	3·2400	1·3416	38	4·243	11	0·5556	31	0·7454	21	0·2357	6	0·2553	24	0·180	1·5136	35
·81	3·2761	·3454	37	·254	12	·5525	30	·7433	21	·2351	7	·2577	24	·181	·5171	34
·82	3·3124	·3491	37	·266	12	·5495	31	·7412	20	·2344	6	·2601	24	·182	·5205	36
·83	3·3489	·3528	37	·278	12	·5464	29	·7392	20	·2338	7	·2625	23	·183	·5241	35
·84	3·3856	·3565	36	·290	11	·5435	30	·7372	20	·2331	6	·2648	24	·184	·5276	35
1·85	3·4225	1·3601	37	4·301	12	0·5405	29	0·7352	20	0·2325	6	0·2672	23	0·185	1·5311	35
·86	3·4596	·3638	37	·313	11	·5376	28	·7332	19	·2319	7	·2695	23	·186	·5346	36
·87	3·4969	·3675	36	·324	12	·5348	29	·7313	20	·2312	6	·2718	24	·187	·5382	35
·88	3·5344	·3711	37	·336	11	·5319	28	·7293	19	·2306	6	·2742	23	·188	·5417	36
·89	3·5721	·3748	36	·347	12	·5291	28	·7274	19	·2300	6	·2765	23	·189	·5453	35
1·90	3·6100	1·3784	36	4·359	11	0·5263	27	0·7255	19	0·2294	6	0·2788	22	0·190	1·5488	36
·91	3·6481	·3820	36	·370	12	·5236	28	·7236	19	·2288	6	·2810	23	·191	·5524	36
·92	3·6864	·3856	36	·382	11	·5208	27	·7217	19	·2282	6	·2833	23	·192	·5560	36
·93	3·7249	·3892	36	·393	12	·5181	26	·7198	18	·2276	6	·2856	22	·193	·5596	35
·94	3·7636	·3928	36	·405	11	·5155	27	·7180	19	·2270	5	·2878	22	·194	·5631	37
1·95	3·8025	1·3964	36	4·416	11	0·5128	26	0·7161	18	0·2265	6	0·2900	23	0·195	1·5668	36
·96	3·8416	·4000	36	·427	11	·5102	26	·7143	18	·2259	6	·2923	22	·196	·5704	36
·97	3·8809	·4036	35	·438	12	·5076	25	·7125	18	·2253	6	·2945	22	·197	·5740	36
·98	3·9204	·4071	36	·450	11	·5051	26	·7107	18	·2247	5	·2967	22	·198	·5776	36
·99	3·9601	·4107	35	·461	11	·5025	25	·7089	18	·2242	6	·2989	21	·199	·5812	37
2·00	4·0000	1·4142		4·472		0·5000		0·7071		0·2236		0·3010		0·200	1·5849	

TABLE 9

x	x^2	$\sqrt{x}$	d	$\sqrt{10x}$	d	$\frac{1}{x}$	d	$\frac{1}{\sqrt{x}}$	d	$\frac{1}{\sqrt{10x}}$	d	$\log x$	d	$\log t$	t	d
2·00	4·0000	1·4142	35	4·472	11	0·5000	25	0·7071	18	0·2236	6	0·3010	22	**0·200**	1·5849	36
·01	4·0401	·4177	36	·483	11	·4975	25	·7053	17	·2230	5	·3032	22	**·201**	·5885	37
·02	4·0804	·4213	35	·494	12	·4950	24	·7036	17	·2225	6	·3054	21	**·202**	·5922	37
·03	4·1209	·4248	35	·506	11	·4926	24	·7019	18	·2219	5	·3075	21	**·203**	·5959	37
·04	4·1616	·4283	35	·517	11	·4902	24	·7001	17	·2214	5	·3096	22	**·204**	·5996	36
2·05	4·2025	1·4318	35	4·528	11	0·4878	24	0·6984	17	0·2209	6	0·3118	21	**0·205**	1·6032	37
·06	4·2436	·4353	34	·539	11	·4854	23	·6967	17	·2203	5	·3139	21	**·206**	·6069	37
·07	4·2849	·4387	35	·550	11	·4831	23	·6950	16	·2198	5	·3160	21	**·207**	·6106	38
·08	4·3264	·4422	35	·561	11	·4808	23	·6934	17	·2193	6	·3181	20	**·208**	·6144	37
·09	4·3681	·4457	34	·572	11	·4785	23	·6917	16	·2187	5	·3201	21	**·209**	·6181	37
2·10	4·4100	1·4491	35	4·583	10	0·4762	23	0·6901	17	0·2182	5	0·3222	21	**0·210**	1·6218	37
·11	4·4521	·4526	34	·593	11	·4739	22	·6884	16	·2177	5	·3243	20	**·211**	·6255	38
·12	4·4944	·4560	35	·604	11	·4717	22	·6868	16	·2172	5	·3263	21	**·212**	·6293	38
·13	4·5369	·4595	34	·615	11	·4695	22	·6852	16	·2167	5	·3284	20	**·213**	·6331	37
·14	4·5796	·4629	34	·626	11	·4673	22	·6836	16	·2162	5	·3304	20	**·214**	·6368	38
2·15	4·6225	1·4663	34	4·637	11	0·4651	21	0·6820	16	0·2157	5	0·3324	21	**0·215**	1·6406	38
·16	4·6656	·4697	34	·648	10	·4630	22	·6804	16	·2152	5	·3345	20	**·216**	·6444	38
·17	4·7089	·4731	34	·658	11	·4608	21	·6788	15	·2147	5	·3365	20	**·217**	·6482	38
·18	4·7524	·4765	34	·669	11	·4587	21	·6773	16	·2142	5	·3385	19	**·218**	·6520	38
·19	4·7961	·4799	33	·680	10	·4566	21	·6757	15	·2137	5	·3404	20	**·219**	·6558	38
2·20	4·8400	1·4832	34	4·690	11	0·4545	20	0·6742	15	0·2132	5	0·3424	20	**0·220**	1·6596	38
·21	4·8841	·4866	34	·701	11	·4525	20	·6727	15	·2127	5	·3444	20	**·221**	·6634	38
·22	4·9284	·4900	33	·712	10	·4505	21	·6712	16	·2122	4	·3464	19	**·222**	·6672	39
·23	4·9729	·4933	34	·722	11	·4484	20	·6696	14	·2118	5	·3483	19	**·223**	·6711	38
·24	5·0176	·4967	33	·733	10	·4464	20	·6682	15	·2113	5	·3502	20	**·224**	·6749	39
2·25	5·0625	1·5000	33	4·743	11	0·4444	19	0·6667	15	0·2108	4	0·3522	19	**0·225**	1·6788	39
·26	5·1076	·5033	34	·754	10	·4425	20	·6652	15	·2104	5	·3541	19	**·226**	·6827	39
·27	5·1529	·5067	33	·764	11	·4405	19	·6637	14	·2099	5	·3560	19	**·227**	·6866	38
·28	5·1984	·5100	33	·775	10	·4386	19	·6623	15	·2094	4	·3579	19	**·228**	·6904	39
·29	5·2441	·5133	33	·785	11	·4367	19	·6608	14	·2090	5	·3598	19	**·229**	·6943	39
2·30	5·2900	1·5166	33	4·796	10	0·4348	19	0·6594	14	0·2085	4	0·3617	19	**0·230**	1·6982	40
·31	5·3361	·5199	33	·806	11	·4329	19	·6580	15	·2081	5	·3636	19	**·231**	·7022	39
·32	5·3824	·5232	32	·817	10	·4310	18	·6565	14	·2076	4	·3655	19	**·232**	·7061	39
·33	5·4289	·5264	33	·827	10	·4292	18	·6551	14	·2072	5	·3674	18	**·233**	·7100	40
·34	5·4756	·5297	33	·837	11	·4274	19	·6537	14	·2067	4	·3692	19	**·234**	·7140	39
2·35	5·5225	1·5330	32	4·848	10	0·4255	18	0·6523	14	0·2063	5	0·3711	18	**0·235**	1·7179	40
·36	5·5696	·5362	33	·858	10	·4237	18	·6509	13	·2058	4	·3729	18	**·236**	·7219	39
·37	5·6169	·5395	32	·868	11	·4219	17	·6496	14	·2054	4	·3747	19	**·237**	·7258	40
·38	5·6644	·5427	33	·879	10	·4202	18	·6482	14	·2050	4	·3766	18	**·238**	·7298	40
·39	5·7121	·5460	32	·889	10	·4184	17	·6468	13	·2046	5	·3784	18	**·239**	·7338	40
2·40	5·7600	1·5492	32	4·899	10	0·4167	18	0·6455	13	0·2041	4	0·3802	18	**0·240**	1·7378	40
·41	5·8081	·5524	32	·909	10	·4149	17	·6442	14	·2037	4	·3820	18	**·241**	·7418	40
·42	5·8564	·5556	32	·919	11	·4132	17	·6428	13	·2033	4	·3838	18	**·242**	·7458	40
·43	5·9049	·5588	32	·930	10	·4115	17	·6415	13	·2029	5	·3856	18	**·243**	·7498	41
·44	5·9536	·5620	32	·940	10	·4098	16	·6402	13	·2024	4	·3874	18	**·244**	·7539	40
2·45	6·0025	1·5652	32	4·950	10	0·4082	17	0·6389	13	0·2020	4	0·3892	17	**0·245**	1·7579	41
·46	6·0516	·5684	32	·960	10	·4065	16	·6376	13	·2016	4	·3909	18	**·246**	·7620	40
·47	6·1009	·5716	32	·970	10	·4049	17	·6363	13	·2012	4	·3927	18	**·247**	·7660	41
·48	6·1504	·5748	32	·980	10	·4032	16	·6350	13	·2008	4	·3945	17	**·248**	·7701	41
·49	6·2001	·5780	31	4·990	10	·4016	16	·6337	12	·2004	4	·3962	17	**·249**	·7742	41
2·50	6·2500	1·5811		5·000		0·4000		0·6325		0·2000		0·3979		**0·250**	1·7783	

TABLE 9

x	x^2	$\sqrt{x}$		$\sqrt{10x}$		$\frac{1}{x}$		$\frac{1}{\sqrt{x}}$		$\frac{1}{\sqrt{10x}}$		$\log x$		$\log t$	t	
2·50	6·2500	1·5811	32	5·000	10	0·4000	16	0·6325	13	0·20000	40	0·3979	18	**0·250**	1·7783	41
·51	6·3001	·5843	32	·010	10	·3984	16	·6312	13	·19960	40	·3997	17	**·251**	·7824	41
·52	6·3504	·5875	31	·020	10	·3968	15	·6299	12	·19920	39	·4014	17	**·252**	·7865	41
·53	6·4009	·5906	31	·030	10	·3953	16	·6287	12	·19881	39	·4031	17	**·253**	·7906	41
·54	6·4516	·5937	32	·040	10	·3937	15	·6275	13	·19842	39	·4048	17	**·254**	·7947	42
2·55	6·5025	1·5969	31	5·050	10	0·3922	16	0·6262	12	0·19803	39	0·4065	17	**0·255**	1·7989	41
·56	6·5536	·6000	31	·060	10	·3906	15	·6250	12	·19764	38	·4082	17	**·256**	·8030	42
·57	6·6049	·6031	31	·070	9	·3891	15	·6238	12	·19726	39	·4099	17	**·257**	·8072	41
·58	6·6564	·6062	31	·079	10	·3876	15	·6226	12	·19687	38	·4116	17	**·258**	·8113	42
·59	6·7081	·6093	32	·089	10	·3861	15	·6214	12	·19649	37	·4133	17	**·259**	·8155	42
2·60	6·7600	1·6125	30	5·099	10	0·3846	15	0·6202	12	0·19612	38	0·4150	16	**0·260**	1·8197	42
·61	6·8121	·6155	31	·109	10	·3831	14	·6190	12	·19574	37	·4166	17	**·261**	·8239	42
·62	6·8644	·6186	31	·119	9	·3817	15	·6178	12	·19537	38	·4183	17	**·262**	·8281	42
·63	6·9169	·6217	31	·128	10	·3802	14	·6166	11	·19499	37	·4200	16	**·263**	·8323	42
·64	6·9696	·6248	31	·138	10	·3788	14	·6155	12	·19462	36	·4216	16	**·264**	·8365	43
2·65	7·0225	1·6279	31	5·148	10	0·3774	15	0·6143	12	0·19426	37	0·4232	17	**0·265**	1·8408	42
·66	7·0756	·6310	30	·158	9	·3759	14	·6131	11	·19389	36	·4249	16	**·266**	·8450	43
·67	7·1289	·6340	31	·167	10	·3745	14	·6120	12	·19353	36	·4265	16	**·267**	·8493	42
·68	7·1824	·6371	30	·177	10	·3731	14	·6108	11	·19317	36	·4281	17	**·268**	·8535	43
·69	7·2361	·6401	31	·187	9	·3717	13	·6097	11	·19281	36	·4298	16	**·269**	·8578	43
2·70	7·2900	1·6432	30	5·196	10	0·3704	14	0·6086	11	0·19245	36	0·4314	16	**0·270**	1·8621	43
·71	7·3441	·6462	30	·206	9	·3690	14	·6075	12	·19209	35	·4330	16	**·271**	·8664	43
·72	7·3984	·6492	31	·215	10	·3676	13	·6063	11	·19174	35	·4346	16	**·272**	·8707	43
·73	7·4529	·6523	30	·225	10	·3663	13	·6052	11	·19139	35	·4362	16	**·273**	·8750	43
·74	7·5076	·6553	30	·235	9	·3650	14	·6041	11	·19104	35	·4378	15	**·274**	·8793	43
2·75	7·5625	1·6583	30	5·244	10	0·3636	13	0·6030	11	0·19069	34	0·4393	16	**0·275**	1·8836	44
·76	7·6176	·6613	30	·254	9	·3623	13	·6019	11	·19035	35	·4409	16	**·276**	·8880	43
·77	7·6729	·6643	30	·263	10	·3610	13	·6008	10	·19000	34	·4425	15	**·277**	·8923	44
·78	7·7284	·6673	30	·273	9	·3597	13	·5998	11	·18966	34	·4440	16	**·278**	·8967	44
·79	7·7841	·6703	30	·282	10	·3584	13	·5987	11	·18932	34	·4456	16	**·279**	·9011	44
2·80	7·8400	1·6733	30	5·292	9	0·3571	12	0·5976	11	0·18898	33	0·4472	15	**0·280**	1·9055	44
·81	7·8961	·6763	30	·301	9	·3559	13	·5965	10	·18865	34	·4487	15	**·281**	·9099	44
·82	7·9524	·6793	30	·310	10	·3546	12	·5955	11	·18831	33	·4502	16	**·282**	·9143	44
·83	8·0089	·6823	29	·320	9	·3534	13	·5944	10	·18798	33	·4518	15	**·283**	·9187	44
·84	8·0656	·6852	30	·329	10	·3521	12	·5934	11	·18765	33	·4533	15	**·284**	·9231	44
2·85	8·1225	1·6882	30	5·339	9	0·3509	12	0·5923	10	0·18732	33	0·4548	16	**0·285**	1·9275	45
·86	8·1796	·6912	29	·348	9	·3497	13	·5913	10	·18699	33	·4564	15	**·286**	·9320	44
·87	8·2369	·6941	30	·357	10	·3484	12	·5903	10	·18666	32	·4579	15	**·287**	·9364	45
·88	8·2944	·6971	29	·367	9	·3472	12	·5893	11	·18634	32	·4594	15	**·288**	·9409	45
·89	8·3521	·7000	29	·376	9	·3460	12	·5882	10	·18602	32	·4609	15	**·289**	·9454	44
2·90	8·4100	1·7029	30	5·385	9	0·3448	12	0·5872	10	0·18570	32	0·4624	15	**0·290**	1·9498	45
·91	8·4681	·7059	29	·394	10	·3436	11	·5862	10	·18538	32	·4639	15	**·291**	·9543	45
·92	8·5264	·7088	29	·404	9	·3425	12	·5852	10	·18506	32	·4654	15	**·292**	·9588	46
·93	8·5849	·7117	29	·413	9	·3413	12	·5842	10	·18474	31	·4669	14	**·293**	·9634	45
·94	8·6436	·7146	30	·422	9	·3401	11	·5832	10	·18443	32	·4683	15	**·294**	·9679	45
2·95	8·7025	1·7176	29	5·431	10	0·3390	12	0·5822	10	0·18411	31	0·4698	15	**0·295**	1·9724	46
·96	8·7616	·7205	29	·441	9	·3378	11	·5812	9	·18380	31	·4713	15	**·296**	·9770	45
·97	8·8209	·7234	29	·450	9	·3367	11	·5803	10	·18349	30	·4728	14	**·297**	·9815	46
·98	8·8804	·7263	29	·459	9	·3356	12	·5793	10	·18319	31	·4742	15	**·298**	·9861	46
·99	8·9401	·7292	29	·468	9	·3344	11	·5783	9	·18288	31	·4757	14	**·299**	·9907	46
3·00	9·0000	1·7321		5·477		0·3333		0·5774		0·18257		0·4771		**0·300**	1·9953	

TABLE 9

x	x^2	$\sqrt{x}$		$\sqrt{10x}$		$\frac{1}{x}$		$\frac{1}{\sqrt{x}}$		$\frac{1}{\sqrt{10x}}$		$\log x$		$\log t$	t	
3·00	9·0000	1·7321	28	5·477	9	0·3333	11	0·5774	10	0·18257	30	0·4771	15	0·300	1·995	5
·01	9·0601	·7349	29	·486	9	·3322	11	·5764	10	·18227	30	·4786	14	·301	2·000	4
·02	9·1204	·7378	29	·495	10	·3311	11	·5754	9	·18197	30	·4800	14	·302	·004	5
·03	9·1809	·7407	29	·505	9	·3300	11	·5745	10	·18167	30	·4814	15	·303	·009	5
·04	9·2416	·7436	28	·514	9	·3289	10	·5735	9	·18137	30	·4829	14	·304	·014	4
3·05	9·3025	1·7464	29	5·523	9	0·3279	11	0·5726	9	0·18107	29	0·4843	14	0·305	2·018	5
·06	9·3636	·7493	28	·532	9	·3268	11	·5717	10	·18078	30	·4857	14	·306	·023	5
·07	9·4249	·7521	29	·541	9	·3257	10	·5707	9	·18048	29	·4871	15	·307	·028	4
·08	9·4864	·7550	28	·550	9	·3247	11	·5698	9	·18019	29	·4886	14	·308	·032	5
·09	9·5481	·7578	29	·559	9	·3236	10	·5689	9	·17990	29	·4900	14	·309	·037	5
3·10	9·6100	1·7607	28	5·568	9	0·3226	11	0·5680	10	0·17961	29	0·4914	14	0·310	2·042	4
·11	9·6721	·7635	29	·577	9	·3215	10	·5670	9	·17932	29	·4928	14	·311	·046	5
·12	9·7344	·7664	28	·586	9	·3205	10	·5661	9	·17903	29	·4942	13	·312	·051	5
·13	9·7969	·7692	28	·595	9	·3195	10	·5652	9	·17874	28	·4955	14	·313	·056	5
·14	9·8596	·7720	28	·604	8	·3185	10	·5643	9	·17846	29	·4969	14	·314	·061	4
3·15	9·9225	1·7748	28	5·612	9	0·3175	10	0·5634	9	0·17817	28	0·4983	14	0·315	2·065	5
·16	9·9856	·7776	28	·621	9	·3165	10	·5625	8	·17789	28	·4997	14	·316	·070	5
·17	10·0489	·7804	29	·630	9	·3155	10	·5617	9	·17761	28	·5011	13	·317	·075	5
·18	10·1124	·7833	28	·639	9	·3145	10	·5608	9	·17733	28	·5024	14	·318	·080	4
·19	10·1761	·7861	28	·648	9	·3135	10	·5599	9	·17705	27	·5038	13	·319	·084	5
3·20	10·2400	1·7889	27	5·657	9	0·3125	10	0·5590	9	0·17678	28	0·5051	14	0·320	2·089	5
·21	10·3041	·7916	28	·666	9	·3115	9	·5581	8	·17650	27	·5065	14	·321	·094	5
·22	10·3684	·7944	28	·675	8	·3106	10	·5573	9	·17623	28	·5079	13	·322	·099	5
·23	10·4329	·7972	28	·683	9	·3096	10	·5564	8	·17595	27	·5092	13	·323	·104	5
·24	10·4976	·8000	28	·692	9	·3086	9	·5556	9	·17568	27	·5105	14	·324	·109	4
3·25	10·5625	1·8028	27	5·701	9	0·3077	10	0·5547	9	0·17541	27	0·5119	13	0·325	2·113	5
·26	10·6276	·8055	28	·710	8	·3067	9	·5538	8	·17514	27	·5132	13	·326	·118	5
·27	10·6929	·8083	28	·718	9	·3058	9	·5530	8	·17487	26	·5145	14	·327	·123	5
·28	10·7584	·8111	27	·727	9	·3049	9	·5522	9	·17461	27	·5159	13	·328	·128	5
·29	10·8241	·8138	28	·736	9	·3040	10	·5513	8	·17434	26	·5172	13	·329	·133	5
3·30	10·8900	1·8166	27	5·745	8	0·3030	9	0·5505	9	0·17408	27	0·5185	13	0·330	2·138	5
·31	10·9561	·8193	28	·753	9	·3021	9	·5496	8	·17381	26	·5198	13	·331	·143	5
·32	11·0224	·8221	27	·762	9	·3012	9	·5488	8	·17355	26	·5211	13	·332	·148	5
·33	11·0889	·8248	28	·771	8	·3003	9	·5480	8	·17329	26	·5224	13	·333	·153	5
·34	11·1556	·8276	27	·779	9	·2994	9	·5472	8	·17303	26	·5237	13	·334	·158	5
3·35	11·2225	1·8303	27	5·788	9	0·2985	9	0·5464	9	0·17277	25	0·5250	13	0·335	2·163	5
·36	11·2896	·8330	28	·797	8	·2976	9	·5455	8	·17252	26	·5263	13	·336	·168	5
·37	11·3569	·8358	27	·805	9	·2967	8	·5447	8	·17226	25	·5276	13	·337	·173	5
·38	11·4244	·8385	27	·814	8	·2959	9	·5439	8	·17201	26	·5289	13	·338	·178	5
·39	11·4921	·8412	27	·822	9	·2950	9	·5431	8	·17175	25	·5302	13	·339	·183	5
3·40	11·5600	1·8439	27	5·831	9	0·2941	8	0·5423	8	0·17150	25	0·5315	13	0·340	2·188	5
·41	11·6281	·8466	27	·840	8	·2933	9	·5415	8	·17125	25	·5328	12	·341	·193	5
·42	11·6964	·8493	27	·848	9	·2924	9	·5407	8	·17100	25	·5340	13	·342	·198	5
·43	11·7649	·8520	27	·857	8	·2915	8	·5399	7	·17075	25	·5353	13	·343	·203	5
·44	11·8336	·8547	27	·865	9	·2907	8	·5392	8	·17050	25	·5366	12	·344	·208	5
3·45	11·9025	1·8574	27	5·874	8	0·2899	9	0·5384	8	0·17025	24	0·5378	13	0·345	2·213	5
·46	11·9716	·8601	27	·882	9	·2890	8	·5376	8	·17001	25	·5391	12	·346	·218	5
·47	12·0409	·8628	27	·891	8	·2882	8	·5368	7	·16976	24	·5403	13	·347	·223	5
·48	12·1104	·8655	27	·899	9	·2874	9	·5361	8	·16952	25	·5416	12	·348	·228	6
·49	12·1801	·8682	26	·908	8	·2865	8	·5353	8	·16927	24	·5428	13	·349	·234	5
3·50	12·2500	1·8708		5·916		0·2857		0·5345		0·16903		0·5441		0·350	2·239	

TABLE 9

x	x^2	$\sqrt{x}$	$\sqrt{10x}$	$\frac{1}{x}$	$\frac{1}{\sqrt{x}}$	$\frac{1}{\sqrt{10x}}$	log x	log t	t
3·50	12·2500	1·8708 27	5·916 9	0·2857 8	0·5345 7	0·16903 24	0·5441 12	0·350	2·239 5
·51	12·3201	·8735 27	·925 8	·2849 8	·5338 8	·16879 24	·5453 12	·351	·244 5
·52	12·3904	·8762 26	·933 8	·2841 8	·5330 8	·16855 24	·5465 13	·352	·249 5
·53	12·4609	·8788 27	·941 9	·2833 8	·5322 7	·16831 24	·5478 12	·353	·254 5
·54	12·5316	·8815 26	·950 8	·2825 8	·5315 8	·16807 23	·5490 12	·354	·259 6
3·55	12·6025	1·8841 27	5·958 9	0·2817 8	0·5307 7	0·16784 24	0·5502 12	0·355	2·265 5
·56	12·6736	·8868 26	·967 8	·2809 8	·5300 7	·16760 23	·5514 13	·356	·270 5
·57	12·7449	·8894 27	·975 8	·2801 8	·5293 8	·16737 24	·5527 12	·357	·275 5
·58	12·8164	·8921 26	·983 9	·2793 7	·5285 7	·16713 23	·5539 12	·358	·280 6
·59	12·8881	·8947 27	5·992 8	·2786 8	·5278 8	·16690 23	·5551 12	·359	·286 5
3·60	12·9600	1·8974 26	6·000 8	0·2778 8	0·5270 7	0·16667 23	0·5563 12	0·360	2·291 5
·61	13·0321	·9000 26	·008 9	·2770 8	·5263 7	·16644 23	·5575 12	·361	·296 5
·62	13·1044	·9026 27	·017 8	·2762 7	·5256 7	·16621 23	·5587 12	·362	·301 6
·63	13·1769	·9053 26	·025 8	·2755 8	·5249 8	·16598 23	·5599 12	·363	·307 5
·64	13·2496	·9079 26	·033 9	·2747 7	·5241 7	·16575 23	·5611 12	·364	·312 5
3·65	13·3225	1·9105 26	6·042 8	0·2740 8	0·5234 7	0·16552 23	0·5623 12	0·365	2·317 6
·66	13·3956	·9131 26	·050 8	·2732 7	·5227 7	·16529 22	·5635 12	·366	·323 5
·67	13·4689	·9157 26	·058 8	·2725 8	·5220 7	·16507 22	·5647 11	·367	·328 5
·68	13·5424	·9183 26	·066 9	·2717 7	·5213 7	·16485 23	·5658 12	·368	·333 6
·69	13·6161	·9209 26	·075 8	·2710 7	·5206 7	·16462 22	·5670 12	·369	·339 5
3·70	13·6900	1·9235 26	6·083 8	0·2703 8	0·5199 7	0·16440 22	0·5682 12	0·370	2·344 6
·71	13·7641	·9261 26	·091 8	·2695 7	·5192 7	·16418 22	·5694 11	·371	·350 5
·72	13·8384	·9287 26	·099 8	·2688 7	·5185 7	·16396 22	·5705 12	·372	·355 5
·73	13·9129	·9313 26	·107 9	·2681 7	·5178 7	·16374 22	·5717 12	·373	·360 6
·74	13·9876	·9339 26	·116 8	·2674 7	·5171 7	·16352 22	·5729 11	·374	·366 5
3·75	14·0625	1·9365 26	6·124 8	0·2667 7	0·5164 7	0·16330 22	0·5740 12	0·375	2·371 6
·76	14·1376	·9391 25	·132 8	·2660 7	·5157 7	·16308 21	·5752 11	·376	·377 5
·77	14·2129	·9416 26	·140 8	·2653 7	·5150 7	·16287 22	·5763 12	·377	·382 6
·78	14·2884	·9442 26	·148 8	·2646 7	·5143 6	·16265 21	·5775 11	·378	·388 5
·79	14·3641	·9468 26	·156 8	·2639 7	·5137 7	·16244 22	·5786 12	·379	·393 6
3·80	14·4400	1·9494 25	6·164 9	0·2632 7	0·5130 7	0·16222 21	0·5798 11	0·380	2·399 5
·81	14·5161	·9519 26	·173 8	·2625 7	·5123 7	·16201 21	·5809 12	·381	·404 6
·82	14·5924	·9545 25	·181 8	·2618 7	·5116 6	·16180 22	·5821 11	·382	·410 5
·83	14·6689	·9570 26	·189 8	·2611 7	·5110 7	·16158 21	·5832 11	·383	·415 6
·84	14·7456	·9596 25	·197 8	·2604 7	·5103 7	·16137 21	·5843 12	·384	·421 6
3·85	14·8225	1·9621 26	6·205 8	0·2597 6	0·5096 6	0·16116 20	0·5855 11	0·385	2·427 5
·86	14·8996	·9647 25	·213 8	·2591 7	·5090 7	·16096 21	·5866 11	·386	·432 6
·87	14·9769	·9672 26	·221 8	·2584 7	·5083 6	·16075 21	·5877 11	·387	·438 5
·88	15·0544	·9698 25	·229 8	·2577 6	·5077 7	·16054 21	·5888 11	·388	·443 6
·89	15·1321	·9723 25	·237 8	·2571 7	·5070 6	·16033 20	·5899 12	·389	·449 6
3·90	15·2100	1·9748 26	6·245 8	0·2564 6	0·5064 7	0·16013 21	0·5911 11	0·390	2·455 5
·91	15·2881	·9774 25	·253 8	·2558 7	·5057 6	·15992 20	·5922 11	·391	·460 6
·92	15·3664	·9799 25	·261 8	·2551 6	·5051 7	·15972 20	·5933 11	·392	·466 6
·93	15·4449	·9824 25	·269 8	·2545 7	·5044 6	·15952 21	·5944 11	·393	·472 5
·94	15·5236	·9849 26	·277 8	·2538 6	·5038 6	·15931 20	·5955 11	·394	·477 6
3·95	15·6025	1·9875 25	6·285 8	0·2532 7	0·5032 7	0·15911 20	0·5966 11	0·395	2·483 6
·96	15·6816	·9900 25	·293 8	·2525 6	·5025 6	·15891 20	·5977 11	·396	·489 6
·97	15·7609	·9925 25	·301 8	·2519 6	·5019 6	·15871 20	·5988 11	·397	·495 5
·98	15·8404	·9950 25	·309 8	·2513 7	·5013 7	·15851 20	·5999 11	·398	·500 6
·99	15·9201	1·9975 25	·317 8	·2506 6	·5006 6	·15831 20	·6010 11	·399	·506 6
4·00	16·0000	2·0000	6·325	0·2500	0·5000	0·15811	0·6021	0·400	2·512

TABLE 9

x	x^2	$\sqrt{x}$	$\sqrt{10x}$	Δ	$\frac{1}{x}$	Δ	$\frac{1}{\sqrt{x}}$	Δ	$\frac{1}{\sqrt{10x}}$	Δ	$\log x$	Δ	$\log t$	t	Δ
4·00	16·0000	2·000	6·325	7	0·2500	6	0·5000	6	0·15811	19	0·6021	10	**0·400**	2·512	6
·01	16·0801	·002	·332	8	·2494	6	·4994	6	·15792	20	·6031	11	**·401**	·518	5
·02	16·1604	·005	·340	8	·2488	7	·4988	7	·15772	20	·6042	11	**·402**	·523	6
·03	16·2409	·007	·348	8	·2481	6	·4981	6	·15752	19	·6053	11	**·403**	·529	6
·04	16·3216	·010	·356	8	·2475	6	·4975	6	·15733	20	·6064	11	**·404**	·535	6
4·05	16·4025	2·012	6·364	8	0·2469	6	0·4969	6	0·15713	19	0·6075	10	**0·405**	2·541	6
·06	16·4836	·015	·372	8	·2463	6	·4963	6	·15694	19	·6085	11	**·406**	·547	6
·07	16·5649	·017	·380	7	·2457	6	·4957	6	·15675	19	·6096	11	**·407**	·553	6
·08	16·6464	·020	·387	8	·2451	6	·4951	6	·15656	20	·6107	10	**·408**	·559	5
·09	16·7281	·022	·395	8	·2445	6	·4945	6	·15636	19	·6117	11	**·409**	·564	6
4·10	16·8100	2·025	6·403	8	0·2439	6	0·4939	6	0·15617	19	0·6128	10	**0·410**	2·570	6
·11	16·8921	·027	·411	8	·2433	6	·4933	6	·15598	19	·6138	11	**·411**	·576	6
·12	16·9744	·030	·419	8	·2427	6	·4927	6	·15579	18	·6149	11	**·412**	·582	6
·13	17·0569	·032	·427	7	·2421	6	·4921	6	·15561	19	·6160	10	**·413**	·588	6
·14	17·1396	·035	·434	8	·2415	5	·4915	6	·15542	19	·6170	10	**·414**	·594	6
4·15	17·2225	2·037	6·442	8	0·2410	6	0·4909	6	0·15523	19	0·6180	11	**0·415**	2·600	6
·16	17·3056	·040	·450	8	·2404	6	·4903	6	·15504	18	·6191	10	**·416**	·606	6
·17	17·3889	·042	·458	7	·2398	6	·4897	6	·15486	19	·6201	11	**·417**	·612	6
·18	17·4724	·045	·465	8	·2392	5	·4891	6	·15467	18	·6212	10	**·418**	·618	6
·19	17·5561	·047	·473	8	·2387	6	·4885	5	·15449	19	·6222	10	**·419**	·624	6
4·20	17·6400	2·049	6·481	7	0·2381	6	0·4880	6	0·15430	18	0·6232	11	**0·420**	2·630	6
·21	17·7241	·052	·488	8	·2375	5	·4874	6	·15412	18	·6243	10	**·421**	·636	6
·22	17·8084	·054	·496	8	·2370	6	·4868	6	·15394	18	·6253	10	**·422**	·642	7
·23	17·8929	·057	·504	8	·2364	6	·4862	6	·15376	19	·6263	11	**·423**	·649	6
·24	17·9776	·059	·512	7	·2358	5	·4856	5	·15357	18	·6274	10	**·424**	·655	6
4·25	18·0625	2·062	6·519	8	0·2353	6	0·4851	6	0·15339	18	0·6284	10	**0·425**	2·661	6
·26	18·1476	·064	·527	8	·2347	5	·4845	6	·15321	18	·6294	10	**·426**	·667	6
·27	18·2329	·066	·535	7	·2342	6	·4839	5	·15303	18	·6304	10	**·427**	·673	6
·28	18·3184	·069	·542	8	·2336	5	·4834	6	·15285	17	·6314	11	**·428**	·679	6
·29	18·4041	·071	·550	7	·2331	5	·4828	6	·15268	18	·6325	10	**·429**	·685	7
4·30	18·4900	2·074	6·557	8	0·2326	6	0·4822	5	0·15250	18	0·6335	10	**0·430**	2·692	6
·31	18·5761	·076	·565	8	·2320	5	·4817	6	·15232	17	·6345	10	**·431**	·698	6
·32	18·6624	·078	·573	7	·2315	6	·4811	5	·15215	18	·6355	10	**·432**	·704	6
·33	18·7489	·081	·580	8	·2309	5	·4806	6	·15197	18	·6365	10	**·433**	·710	6
·34	18·8356	·083	·588	7	·2304	5	·4800	5	·15179	17	·6375	10	**·434**	·716	7
4·35	18·9225	2·086	6·595	8	0·2299	5	0·4795	6	0·15162	17	0·6385	10	**0·435**	2·723	6
·36	19·0096	·088	·603	8	·2294	6	·4789	5	·15145	18	·6395	10	**·436**	·729	6
·37	19·0969	·090	·611	7	·2288	5	·4784	6	·15127	17	·6405	10	**·437**	·735	7
·38	19·1844	·093	·618	8	·2283	5	·4778	5	·15110	17	·6415	10	**·438**	·742	6
·39	19·2721	·095	·626	7	·2278	5	·4773	6	·15093	17	·6425	10	**·439**	·748	6
4·40	19·3600	2·098	6·633	8	0·2273	5	0·4767	5	0·15076	18	0·6435	9	**0·440**	2·754	7
·41	19·4481	·100	·641	7	·2268	6	·4762	5	·15058	17	·6444	10	**·441**	·761	6
·42	19·5364	·102	·648	8	·2262	5	·4757	6	·15041	17	·6454	10	**·442**	·767	6
·43	19·6249	·105	·656	7	·2257	5	·4751	5	·15024	16	·6464	10	**·443**	·773	7
·44	19·7136	·107	·663	8	·2252	5	·4746	6	·15008	17	·6474	10	**·444**	·780	6
4·45	19·8025	2·110	6·671	7	0·2247	5	0·4740	5	0·14991	17	0·6484	9	**0·445**	2·786	7
·46	19·8916	·112	·678	8	·2242	5	·4735	5	·14974	17	·6493	10	**·446**	·793	6
·47	19·9809	·114	·686	7	·2237	5	·4730	5	·14957	17	·6503	10	**·447**	·799	6
·48	20·0704	·117	·693	8	·2232	5	·4725	6	·14940	16	·6513	9	**·448**	·805	7
·49	20·1601	·119	·701	7	·2227	5	·4719	5	·14924	17	·6522	10	**·449**	·812	6
4·50	20·2500	2·121	6·708		0·2222		0·4714		0·14907		0·6532		**0·450**	2·818	

TABLE 9

x	x^2	$\sqrt{x}$	$\sqrt{10x}$		$\frac{1}{x}$		$\frac{1}{\sqrt{x}}$		$\frac{1}{\sqrt{10x}}$		$\log x$		$\log t$	t	
4·50	20·2500	2·121	6·708	8	0·2222	5	0·4714	5	0·14907	16	0·6532	10	0·450	2·818	7
·51	20·3401	·124	·716	7	·2217	5	·4709	5	·14891	17	·6542	9	·451	·825	6
·52	20·4304	·126	·723	8	·2212	4	·4704	6	·14874	16	·6551	10	·452	·831	7
·53	20·5209	·128	·731	7	·2208	5	·4698	5	·14858	17	·6561	10	·453	·838	6
·54	20·6116	·131	·738	7	·2203	5	·4693	5	·14841	16	·6571	9	·454	·844	7
4·55	20·7025	2·133	6·745	8	0·2198	5	0·4688	5	0·14825	16	0·6580	10	0·455	2·851	7
·56	20·7936	·135	·753	7	·2193	5	·4683	5	·14809	16	·6590	9	·456	·858	6
·57	20·8849	·138	·760	8	·2188	5	·4678	5	·14793	17	·6599	10	·457	·864	7
·58	20·9764	·140	·768	7	·2183	4	·4673	5	·14776	16	·6609	9	·458	·871	6
·59	21·0681	·142	·775	7	·2179	5	·4668	5	·14760	16	·6618	10	·459	·877	7
4·60	21·1600	2·145	6·782	8	0·2174	5	0·4663	6	0·14744	16	0·6628	9	0·460	2·884	7
·61	21·2521	·147	·790	7	·2169	4	·4657	5	·14728	16	·6637	9	·461	·891	6
·62	21·3444	·149	·797	7	·2165	5	·4652	5	·14712	16	·6646	10	·462	·897	7
·63	21·4369	·152	·804	8	·2160	5	·4647	5	·14696	15	·6656	9	·463	·904	7
·64	21·5296	·154	·812	7	·2155	4	·4642	5	·14681	16	·6665	10	·464	·911	6
4·65	21·6225	2·156	6·819	7	0·2151	5	0·4637	5	0·14665	16	0·6675	9	0·465	2·917	7
·66	21·7156	·159	·826	8	·2146	5	·4632	5	·14649	16	·6684	9	·466	·924	7
·67	21·8089	·161	·834	7	·2141	4	·4627	4	·14633	15	·6693	9	·467	·931	7
·68	21·9024	·163	·841	7	·2137	5	·4623	5	·14618	16	·6702	10	·468	·938	6
·69	21·9961	·166	·848	8	·2132	4	·4618	5	·14602	16	·6712	9	·469	·944	7
4·70	22·0900	2·168	6·856	7	0·2128	5	0·4613	5	0·14586	15	0·6721	9	0·470	2·951	7
·71	22·1841	·170	·863	7	·2123	4	·4608	5	·14571	15	·6730	9	·471	·958	7
·72	22·2784	·173	·870	7	·2119	5	·4603	5	·14556	16	·6739	10	·472	·965	7
·73	22·3729	·175	·877	8	·2114	4	·4598	5	·14540	15	·6749	9	·473	·972	7
·74	22·4676	·177	·885	7	·2110	5	·4593	5	·14525	15	·6758	9	·474	·979	6
4·75	22·5625	2·179	6·892	7	0·2105	4	0·4588	5	0·14510	16	0·6767	9	0·475	2·985	7
·76	22·6576	·182	·899	8	·2101	5	·4583	4	·14494	15	·6776	9	·476	·992	7
·77	22·7529	·184	·907	7	·2096	4	·4579	5	·14479	15	·6785	9	·477	2·999	7
·78	22·8484	·186	·914	7	·2092	4	·4574	5	·14464	15	·6794	9	·478	3·006	7
·79	22·9441	·189	·921	7	·2088	5	·4569	5	·14449	15	·6803	9	·479	·013	7
4·80	23·0400	2·191	6·928	7	0·2083	4	0·4564	4	0·14434	15	0·6812	9	0·480	3·020	7
·81	23·1361	·193	·935	8	·2079	4	·4560	5	·14419	15	·6821	9	·481	·027	7
·82	23·2324	·195	·943	7	·2075	5	·4555	5	·14404	15	·6830	9	·482	·034	7
·83	23·3289	·198	·950	7	·2070	4	·4550	5	·14389	15	·6839	9	·483	·041	7
·84	23·4256	·200	·957	7	·2066	4	·4545	4	·14374	15	·6848	9	·484	·048	7
4·85	23·5225	2·202	6·964	7	0·2062	4	0·4541	5	0·14359	15	0·6857	9	0·485	3·055	7
·86	23·6196	·205	·971	8	·2058	5	·4536	5	·14344	14	·6866	9	·486	·062	7
·87	23·7169	·207	·979	7	·2053	4	·4531	4	·14330	15	·6875	9	·487	·069	7
·88	23·8144	·209	·986	7	·2049	4	·4527	5	·14315	15	·6884	9	·488	·076	7
·89	23·9121	·211	6·993	7	·2045	4	·4522	4	·14300	14	·6893	9	·489	·083	7
4·90	24·0100	2·214	7·000	7	0·2041	4	0·4518	5	0·14286	15	0·6902	9	0·490	3·090	7
·91	24·1081	·216	·007	7	·2037	4	·4513	5	·14271	14	·6911	9	·491	·097	8
·92	24·2064	·218	·014	7	·2033	5	·4508	4	·14257	15	·6920	8	·492	·105	7
·93	24·3049	·220	·021	8	·2028	4	·4504	5	·14242	14	·6928	9	·493	·112	7
·94	24·4036	·223	·029	7	·2024	4	·4499	4	·14228	15	·6937	9	·494	·119	7
4·95	24·5025	2·225	7·036	7	0·2020	4	0·4495	5	0·14213	14	0·6946	9	0·495	3·126	7
·96	24·6016	·227	·043	7	·2016	4	·4490	4	·14199	14	·6955	9	·496	·133	8
·97	24·7009	·229	·050	7	·2012	4	·4486	5	·14185	14	·6964	8	·497	·141	7
·98	24·8004	·232	·057	7	·2008	4	·4481	4	·14171	15	·6972	9	·498	·148	7
·99	24·9001	·234	·064	7	·2004	4	·4477	5	·14156	14	·6981	9	·499	·155	7
5·00	25·0000	2·236	7·071		0·2000		0·4472		0·14142		0·6990		0·500	3·162	

TABLE 9

x	x^2	$\sqrt{x}$	$\sqrt{10x}$	Δ	$\frac{1}{x}$	Δ	$\frac{1}{\sqrt{x}}$	Δ	$\frac{1}{\sqrt{10x}}$	Δ	$\log x$	Δ	$\log t$	t	Δ
5·00	25·0000	2·236	7·071	7	0·20000	40	0·4472	4	0·14142	14	0·6990	8	0·500	3·162	8
·01	25·1001	·238	·078	7	·19960	40	·4468	5	·14128	14	·6998	9	·501	·170	7
·02	25·2004	·241	·085	7	·19920	39	·4463	4	·14114	14	·7007	9	·502	·177	7
·03	25·3009	·243	·092	7	·19881	40	·4459	5	·14100	14	·7016	8	·503	·184	8
·04	25·4016	·245	·099	7	·19841	39	·4454	4	·14086	14	·7024	9	·504	·192	7
5·05	25·5025	2·247	7·106	7	0·19802	39	0·4450	4	0·14072	14	0·7033	9	0·505	3·199	7
·06	25·6036	·249	·113	7	·19763	39	·4446	5	·14058	14	·7042	8	·506	·206	8
·07	25·7049	·252	·120	7	·19724	39	·4441	4	·14044	14	·7050	9	·507	·214	7
·08	25·8064	·254	·127	7	·19685	39	·4437	5	·14030	13	·7059	8	·508	·221	7
·09	25·9081	·256	·134	7	·19646	38	·4432	4	·14017	14	·7067	9	·509	·228	8
5·10	26·0100	2·258	7·141	7	0·19608	39	0·4428	4	0·14003	14	0·7076	8	0·510	3·236	7
·11	26·1121	·261	·148	7	·19569	38	·4424	5	·13989	14	·7084	9	·511	·243	8
·12	26·2144	·263	·155	7	·19531	38	·4419	4	·13975	13	·7093	8	·512	·251	7
·13	26·3169	·265	·162	7	·19493	38	·4415	4	·13962	14	·7101	9	·513	·258	8
·14	26·4196	·267	·169	7	·19455	38	·4411	4	·13948	13	·7110	8	·514	·266	7
5·15	26·5225	2·269	7·176	7	0·19417	37	0·4407	5	0·13935	14	0·7118	8	0·515	3·273	8
·16	26·6256	·272	·183	7	·19380	38	·4402	4	·13921	13	·7126	9	·516	·281	8
·17	26·7289	·274	·190	7	·19342	37	·4398	4	·13908	14	·7135	8	·517	·289	7
·18	26·8324	·276	·197	7	·19305	37	·4394	4	·13894	13	·7143	9	·518	·296	8
·19	26·9361	·278	·204	7	·19268	37	·4390	5	·13881	13	·7152	8	·519	·304	7
5·20	27·0400	2·280	7·211	7	0·19231	37	0·4385	4	0·13868	14	0·7160	8	0·520	3·311	8
·21	27·1441	·283	·218	7	·19194	37	·4381	4	·13854	13	·7168	9	·521	·319	8
·22	27·2484	·285	·225	7	·19157	37	·4377	4	·13841	13	·7177	8	·522	·327	7
·23	27·3529	·287	·232	7	·19120	36	·4373	4	·13828	14	·7185	8	·523	·334	8
·24	27·4576	·289	·239	7	·19084	36	·4369	5	·13814	13	·7193	9	·524	·342	8
5·25	27·5625	2·291	7·246	7	0·19048	37	0·4364	4	0·13801	13	0·7202	8	0·525	3·350	7
·26	27·6676	·293	·253	6	·19011	36	·4360	4	·13788	13	·7210	8	·526	·357	8
·27	27·7729	·296	·259	7	·18975	36	·4356	4	·13775	13	·7218	8	·527	·365	8
·28	27·8784	·298	·266	7	·18939	35	·4352	4	·13762	13	·7226	9	·528	·373	8
·29	27·9841	·300	·273	7	·18904	36	·4348	4	·13749	13	·7235	8	·529	·381	7
5·30	28·0900	2·302	7·280	7	0·18868	36	0·4344	4	0·13736	13	0·7243	8	0·530	3·388	8
·31	28·1961	·304	·287	7	·18832	35	·4340	4	·13723	13	·7251	8	·531	·396	8
·32	28·3024	·307	·294	7	·18797	35	·4336	5	·13710	13	·7259	8	·532	·404	8
·33	28·4089	·309	·301	7	·18762	35	·4331	4	·13697	12	·7267	8	·533	·412	8
·34	28·5156	·311	·308	6	·18727	35	·4327	4	·13685	13	·7275	9	·534	·420	8
5·35	28·6225	2·313	7·314	7	0·18692	35	0·4323	4	0·13672	13	0·7284	8	0·535	3·428	8
·36	28·7296	·315	·321	7	·18657	35	·4319	4	·13659	13	·7292	8	·536	·436	7
·37	28·8369	·317	·328	7	·18622	35	·4315	4	·13646	12	·7300	8	·537	·443	8
·38	28·9444	·319	·335	7	·18587	34	·4311	4	·13634	13	·7308	8	·538	·451	8
·39	29·0521	·322	·342	6	·18553	34	·4307	4	·13621	13	·7316	8	·539	·459	8
5·40	29·1600	2·324	7·348	7	0·18519	35	0·4303	4	0·13608	12	0·7324	8	0·540	3·467	8
·41	29·2681	·326	·355	7	·18484	34	·4299	4	·13596	13	·7332	8	·541	·475	8
·42	29·3764	·328	·362	7	·18450	34	·4295	4	·13583	12	·7340	8	·542	·483	8
·43	29·4849	·330	·369	7	·18416	34	·4291	4	·13571	13	·7348	8	·543	·491	8
·44	29·5936	·332	·376	6	·18382	33	·4287	3	·13558	12	·7356	8	·544	·499	9
5·45	29·7025	2·335	7·382	7	0·18349	34	0·4284	4	0·13546	13	0·7364	8	0·545	3·508	8
·46	29·8116	·337	·389	7	·18315	33	·4280	4	·13533	12	·7372	8	·546	·516	8
·47	29·9209	·339	·396	7	·18282	34	·4276	4	·13521	12	·7380	8	·547	·524	8
·48	30·0304	·341	·403	6	·18248	33	·4272	4	·13509	13	·7388	8	·548	·532	8
·49	30·1401	·343	·409	7	·18215	33	·4268	4	·13496	12	·7396	8	·549	·540	8
5·50	30·2500	2·345	7·416		0·18182		0·4264		0·13484		0·7404		0·550	3·548	

TABLE 9

x	x^2	$\sqrt{x}$	$\sqrt{10x}$		$\frac{1}{x}$		$\frac{1}{\sqrt{x}}$		$\frac{1}{\sqrt{10x}}$		$\log x$		$\log t$	t	
5·50	30·2500	2·345	7·416	7	0·18182	33	0·4264	4	0·13484	12	0·7404	8	0·550	3·548	8
·51	30·3601	·347	·423	7	·18149	33	·4260	4	·13472	12	·7412	7	·551	·556	9
·52	30·4704	·349	·430	6	·18116	33	·4256	4	·13460	13	·7419	8	·552	·565	8
·53	30·5809	·352	·436	7	·18083	32	·4252	3	·13447	12	·7427	8	·553	·573	8
·54	30·6916	·354	·443	7	·18051	33	·4249	4	·13435	12	·7435	8	·554	·581	8
5·55	30·8025	2·356	7·450	7	0·18018	32	0·4245	4	0·13423	12	0·7443	8	0·555	3·589	8
·56	30·9136	·358	·457	6	·17986	33	·4241	4	·13411	12	·7451	8	·556	·597	9
·57	31·0249	·360	·463	7	·17953	32	·4237	4	·13399	12	·7459	7	·557	·606	8
·58	31·1364	·362	·470	7	·17921	32	·4233	3	·13387	12	·7466	8	·558	·614	8
·59	31·2481	·364	·477	6	·17889	32	·4230	4	·13375	12	·7474	8	·559	·622	9
5·60	31·3600	2·366	7·483	7	0·17857	32	0·4226	4	0·13363	12	0·7482	8	0·560	3·631	8
·61	31·4721	·369	·490	7	·17825	31	·4222	4	·13351	12	·7490	7	·561	·639	9
·62	31·5844	·371	·497	6	·17794	32	·4218	4	·13339	12	·7497	8	·562	·648	8
·63	31·6969	·373	·503	7	·17762	32	·4214	3	·13327	11	·7505	8	·563	·656	8
·64	31·8096	·375	·510	7	·17730	31	·4211	4	·13316	12	·7513	7	·564	·664	9
5·65	31·9225	2·377	7·517	6	0·17699	31	0·4207	4	0·13304	12	0·7520	8	0·565	3·673	8
·66	32·0356	·379	·523	7	·17668	31	·4203	3	·13292	12	·7528	8	·566	·681	9
·67	32·1489	·381	·530	7	·17637	31	·4200	4	·13280	11	·7536	7	·567	·690	8
·68	32·2624	·383	·537	6	·17606	31	·4196	4	·13269	12	·7543	8	·568	·698	9
·69	32·3761	·385	·543	7	·17575	31	·4192	3	·13257	12	·7551	8	·569	·707	8
5·70	32·4900	2·387	7·550	6	0·17544	31	0·4189	4	0·13245	11	0·7559	7	0·570	3·715	9
·71	32·6041	·390	·556	7	·17513	30	·4185	4	·13234	12	·7566	8	·571	·724	9
·72	32·7184	·392	·563	7	·17483	31	·4181	3	·13222	11	·7574	8	·572	·733	8
·73	32·8329	·394	·570	6	·17452	30	·4178	4	·13211	12	·7582	7	·573	·741	9
·74	32·9476	·396	·576	7	·17422	31	·4174	4	·13199	11	·7589	8	·574	·750	8
5·75	33·0625	2·398	7·583	6	0·17391	30	0·4170	3	0·13188	12	0·7597	7	0·575	3·758	9
·76	33·1776	·400	·589	7	·17361	30	·4167	4	·13176	11	·7604	8	·576	·767	9
·77	33·2929	·402	·596	7	·17331	30	·4163	4	·13165	12	·7612	7	·577	·776	8
·78	33·4084	·404	·603	6	·17301	30	·4159	3	·13153	11	·7619	8	·578	·784	9
·79	33·5241	·406	·609	7	·17271	30	·4156	4	·13142	11	·7627	7	·579	·793	9
5·80	33·6400	2·408	7·616	6	0·17241	29	0·4152	3	0·13131	12	0·7634	8	0·580	3·802	9
·81	33·7561	·410	·622	7	·17212	30	·4149	4	·13119	11	·7642	7	·581	·811	8
·82	33·8724	·412	·629	6	·17182	29	·4145	3	·13108	11	·7649	8	·582	·819	9
·83	33·9889	·415	·635	7	·17153	30	·4142	4	·13097	11	·7657	7	·583	·828	9
·84	34·1056	·417	·642	7	·17123	29	·4138	4	·13086	12	·7664	8	·584	·837	9
5·85	34·2225	2·419	7·649	6	0·17094	29	0·4134	3	0·13074	11	0·7672	7	0·585	3·846	9
·86	34·3396	·421	·655	7	·17065	29	·4131	4	·13063	11	·7679	7	·586	·855	9
·87	34·4569	·423	·662	6	·17036	29	·4127	3	·13052	11	·7686	8	·587	·864	9
·88	34·5744	·425	·668	7	·17007	29	·4124	4	·13041	11	·7694	7	·588	·873	9
·89	34·6921	·427	·675	6	·16978	29	·4120	3	·13030	11	·7701	8	·589	·882	8
5·90	34·8100	2·429	7·681	7	0·16949	29	0·4117	4	0·13019	11	0·7709	7	0·590	3·890	9
·91	34·9281	·431	·688	6	·16920	28	·4113	3	·13008	11	·7716	7	·591	·899	9
·92	35·0464	·433	·694	7	·16892	29	·4110	3	·12997	11	·7723	8	·592	·908	9
·93	35·1649	·435	·701	6	·16863	28	·4107	4	·12986	11	·7731	7	·593	·917	9
·94	35·2836	·437	·707	7	·16835	28	·4103	3	·12975	11	·7738	7	·594	·926	10
5·95	35·4025	2·439	7·714	6	0·16807	28	0·4100	4	0·12964	11	0·7745	7	0·595	3·936	9
·96	35·5216	·441	·720	7	·16779	29	·4096	3	·12953	11	·7752	8	·596	·945	9
·97	35·6409	·443	·727	6	·16750	28	·4093	4	·12942	10	·7760	7	·597	·954	9
·98	35·7604	·445	·733	7	·16722	28	·4089	3	·12932	11	·7767	7	·598	·963	9
·99	35·8801	·447	·740	6	·16694	27	·4086	4	·12921	11	·7774	8	·599	·972	9
6·00	36·0000	2·449	7·746		0·16667		0·4082		0·12910		0·7782		0·600	3·981	

TABLE 9

x	x^2	$\sqrt{x}$	$\sqrt{10x}$		$\frac{1}{x}$		$\frac{1}{\sqrt{x}}$	$\frac{1}{\sqrt{10x}}$		$\log x$		$\log t$	t	
6·00	36·0000	2·449	7·746	6	0·16667	28	0·4082	0·12910	11	0·7782	7	0·600	3·981	9
·01	36·1201	·452	·752	7	·16639	28	·4079	·12899	11	·7789	7	·601	·990	9
·02	36·2404	·454	·759	6	·16611	27	·4076	·12888	10	·7796	7	·602	3·999	10
·03	36·3609	·456	·765	7	·16584	28	·4072	·12878	11	·7803	7	·603	4·009	9
·04	36·4816	·458	·772	6	·16556	27	·4069	·12867	11	·7810	8	·604	·018	9
6·05	36·6025	2·460	7·778	7	0·16529	27	0·4066	0·12856	10	0·7818	7	0·605	4·027	9
·06	36·7236	·462	·785	6	·16502	28	·4062	·12846	11	·7825	7	·606	·036	10
·07	36·8449	·464	·791	6	·16474	27	·4059	·12835	10	·7832	7	·607	·046	9
·08	36·9664	·466	·797	7	·16447	27	·4056	·12825	11	·7839	7	·608	·055	9
·09	37·0881	·468	·804	6	·16420	27	·4052	·12814	10	·7846	7	·609	·064	10
6·10	37·2100	2·470	7·810	7	0·16393	26	0·4049	0·12804	11	0·7853	7	0·610	4·074	9
·11	37·3321	·472	·817	6	·16367	27	·4046	·12793	10	·7860	8	·611	·083	10
·12	37·4544	·474	·823	6	·16340	27	·4042	·12783	11	·7868	7	·612	·093	9
·13	37·5769	·476	·829	7	·16313	26	·4039	·12772	10	·7875	7	·613	·102	9
·14	37·6996	·478	·836	6	·16287	27	·4036	·12762	10	·7882	7	·614	·111	10
6·15	37·8225	2·480	7·842	7	0·16260	26	0·4032	0·12752	11	0·7889	7	0·615	4·121	9
·16	37·9456	·482	·849	6	·16234	27	·4029	·12741	10	·7896	7	·616	·130	10
·17	38·0689	·484	·855	6	·16207	26	·4026	·12731	10	·7903	7	·617	·140	10
·18	38·1924	·486	·861	7	·16181	26	·4023	·12721	11	·7910	7	·618	·150	9
·19	38·3161	·488	·868	6	·16155	26	·4019	·12710	10	·7917	7	·619	·159	10
6·20	38·4400	2·490	7·874	6	0·16129	26	0·4016	0·12700	10	0·7924	7	0·620	4·169	9
·21	38·5641	·492	·880	7	·16103	26	·4013	·12690	10	·7931	7	·621	·178	10
·22	38·6884	·494	·887	6	·16077	26	·4010	·12680	11	·7938	7	·622	·188	10
·23	38·8129	·496	·893	6	·16051	25	·4006	·12669	10	·7945	7	·623	·198	9
·24	38·9376	·498	·899	7	·16026	26	·4003	·12659	10	·7952	7	·624	·207	10
6·25	39·0625	2·500	7·906	6	0·16000	26	0·4000	0·12649	10	0·7959	7	0·625	4·217	10
·26	39·1876	·502	·912	6	·15974	25	·3997	·12639	10	·7966	7	·626	·227	9
·27	39·3129	·504	·918	7	·15949	25	·3994	·12629	10	·7973	7	·627	·236	10
·28	39·4384	·506	·925	6	·15924	26	·3990	·12619	10	·7980	7	·628	·246	10
·29	39·5641	·508	·931	6	·15898	25	·3987	·12609	10	·7987	6	·629	·256	10
6·30	39·6900	2·510	7·937	7	0·15873	25	0·3984	0·12599	10	0·7993	7	0·630	4·266	10
·31	39·8161	·512	·944	6	·15848	25	·3981	·12589	10	·8000	7	·631	·276	9
·32	39·9424	·514	·950	6	·15823	25	·3978	·12579	10	·8007	7	·632	·285	10
·33	40·0689	·516	·956	6	·15798	25	·3975	·12569	10	·8014	7	·633	·295	10
·34	40·1956	·518	·962	7	·15773	25	·3972	·12559	10	·8021	7	·634	·305	10
6·35	40·3225	2·520	7·969	6	0·15748	25	0·3968	0·12549	10	0·8028	7	0·635	4·315	10
·36	40·4496	·522	·975	6	·15723	24	·3965	·12539	10	·8035	6	·636	·325	10
·37	40·5769	·524	·981	6	·15699	25	·3962	·12529	9	·8041	7	·637	·335	10
·38	40·7044	·526	·987	7	·15674	25	·3959	·12520	10	·8048	7	·638	·345	10
·39	40·8321	·528	7·994	6	·15649	24	·3956	·12510	10	·8055	7	·639	·355	10
6·40	40·9600	2·530	8·000	6	0·15625	24	0·3953	0·12500	10	0·8062	7	0·640	4·365	10
·41	41·0881	·532	·006	6	·15601	25	·3950	·12490	9	·8069	6	·641	·375	10
·42	41·2164	·534	·012	7	·15576	24	·3947	·12481	10	·8075	7	·642	·385	10
·43	41·3449	·536	·019	6	·15552	24	·3944	·12471	10	·8082	7	·643	·395	11
·44	41·4736	·538	·025	6	·15528	24	·3941	·12461	10	·8089	7	·644	·406	10
6·45	41·6025	2·540	8·031	6	0·15504	24	0·3937	0·12451	9	0·8096	6	0·645	4·416	10
·46	41·7316	·542	·037	7	·15480	24	·3934	·12442	10	·8102	7	·646	·426	10
·47	41·8609	·544	·044	6	·15456	24	·3931	·12432	9	·8109	7	·647	·436	10
·48	41·9904	·546	·050	6	·15432	24	·3928	·12423	10	·8116	6	·648	·446	11
·49	42·1201	·548	·056	6	·15408	23	·3925	·12413	10	·8122	7	·649	·457	10
6·50	42·2500	2·550	8·062		0·15385		0·3922	0·12403		0·8129		0·650	4·467	

TABLE 9

x	x^2	$\sqrt{x}$	$\sqrt{10x}$		$\frac{1}{x}$		$\frac{1}{\sqrt{x}}$	$\frac{1}{\sqrt{10x}}$		log x		log t	t	
6·50	42·2500	2·550	8·062	6	0·15385	24	0·3922	0·12403	9	0·8129	7	**0·650**	4·467	10
·51	42·3801	·551	·068	7	·15361	24	·3919	·12394	10	·8136	6	**·651**	·477	10
·52	42·5104	·553	·075	6	·15337	23	·3916	·12384	9	·8142	7	**·652**	·487	11
·53	42·6409	·555	·081	6	·15314	23	·3913	·12375	10	·8149	7	**·653**	·498	10
·54	42·7716	·557	·087	6	·15291	24	·3910	·12365	9	·8156	6	**·654**	·508	11
6·55	42·9025	2·559	8·093	6	0·15267	23	0·3907	0·12356	9	0·8162	7	**0·655**	4·519	10
·56	43·0336	·561	·099	7	·15244	23	·3904	·12347	10	·8169	7	**·656**	·529	10
·57	43·1649	·563	·106	6	·15221	23	·3901	·12337	9	·8176	6	**·657**	·539	11
·58	43·2964	·565	·112	6	·15198	23	·3898	·12328	10	·8182	7	**·658**	·550	10
·59	43·4281	·567	·118	6	·15175	23	·3895	·12318	9	·8189	6	**·659**	·560	11
6·60	43·5600	2·569	8·124	6	0·15152	23	0·3892	0·12309	9	0·8195	7	**0·660**	4·571	10
·61	43·6921	·571	·130	6	·15129	23	·3890	·12300	9	·8202	7	**·661**	·581	11
·62	43·8244	·573	·136	6	·15106	23	·3887	·12291	10	·8209	6	**·662**	·592	11
·63	43·9569	·575	·142	7	·15083	23	·3884	·12281	9	·8215	7	**·663**	·603	10
·64	44·0896	·577	·149	6	·15060	22	·3881	·12272	9	·8222	6	**·664**	·613	11
6·65	44·2225	2·579	8·155	6	0·15038	23	0·3878	0·12263	9	0·8228	7	**0·665**	4·624	10
·66	44·3556	·581	·161	6	·15015	22	·3875	·12254	10	·8235	6	**·666**	·634	11
·67	44·4889	·583	·167	6	·14993	23	·3872	·12244	9	·8241	7	**·667**	·645	11
·68	44·6224	·585	·173	6	·14970	22	·3869	·12235	9	·8248	6	**·668**	·656	11
·69	44·7561	·587	·179	6	·14948	23	·3866	·12226	9	·8254	7	**·669**	·667	10
6·70	44·8900	2·588	8·185	6	0·14925	22	0·3863	0·12217	9	0·8261	6	**0·670**	4·677	11
·71	45·0241	·590	·191	7	·14903	22	·3860	·12208	9	·8267	7	**·671**	·688	11
·72	45·1584	·592	·198	6	·14881	22	·3858	·12199	9	·8274	6	**·672**	·699	11
·73	45·2929	·594	·204	6	·14859	22	·3855	·12190	9	·8280	7	**·673**	·710	11
·74	45·4276	·596	·210	6	·14837	22	·3852	·12181	9	·8287	6	**·674**	·721	11
6·75	45·5625	2·598	8·216	6	0·14815	22	0·3849	0·12172	9	0·8293	6	**0·675**	4·732	10
·76	45·6976	·600	·222	6	·14793	22	·3846	·12163	9	·8299	7	**·676**	·742	11
·77	45·8329	·602	·228	6	·14771	22	·3843	·12154	9	·8306	6	**·677**	·753	11
·78	45·9684	·604	·234	6	·14749	21	·3840	·12145	9	·8312	7	**·678**	·764	11
·79	46·1041	·606	·240	6	·14728	22	·3838	·12136	9	·8319	6	**·679**	·775	11
6·80	46·2400	2·608	8·246	6	0·14706	22	0·3835	0·12127	9	0·8325	6	**0·680**	4·786	11
·81	46·3761	·610	·252	6	·14684	21	·3832	·12118	9	·8331	7	**·681**	·797	11
·82	46·5124	·612	·258	6	·14663	22	·3829	·12109	9	·8338	6	**·682**	·808	11
·83	46·6489	·613	·264	6	·14641	21	·3826	·12100	9	·8344	7	**·683**	·819	12
·84	46·7856	·615	·270	6	·14620	21	·3824	·12091	9	·8351	6	**·684**	·831	11
6·85	46·9225	2·617	8·276	7	0·14599	22	0·3821	0·12082	8	0·8357	6	**0·685**	4·842	11
·86	47·0596	·619	·283	6	·14577	21	·3818	·12074	9	·8363	7	**·686**	·853	11
·87	47·1969	·621	·289	6	·14556	21	·3815	·12065	9	·8370	6	**·687**	·864	11
·88	47·3344	·623	·295	6	·14535	21	·3812	·12056	9	·8376	6	**·688**	·875	12
·89	47·4721	·625	·301	6	·14514	21	·3810	·12047	8	·8382	6	**·689**	·887	11
6·90	47·6100	2·627	8·307	6	0·14493	21	0·3807	0·12039	9	0·8388	7	**0·690**	4·898	11
·91	47·7481	·629	·313	6	·14472	21	·3804	·12030	9	·8395	6	**·691**	·909	11
·92	47·8864	·631	·319	6	·14451	21	·3801	·12021	9	·8401	6	**·692**	·920	12
·93	48·0249	·632	·325	6	·14430	21	·3799	·12012	8	·8407	7	**·693**	·932	11
·94	48·1636	·634	·331	6	·14409	21	·3796	·12004	9	·8414	6	**·694**	·943	12
6·95	48·3025	2·636	8·337	6	0·14388	20	0·3793	0·11995	8	0·8420	6	**0·695**	4·955	11
·96	48·4416	·638	·343	6	·14368	21	·3790	·11987	9	·8426	6	**·696**	·966	11
·97	48·5809	·640	·349	6	·14347	20	·3788	·11978	9	·8432	7	**·697**	·977	12
·98	48·7204	·642	·355	6	·14327	21	·3785	·11969	8	·8439	6	**·698**	4·989	11
·99	48·8601	·644	·361	6	·14306	20	·3782	·11961	9	·8445	6	**·699**	5·000	12
7·00	49·0000	2·646	8·367		0·14286		0·3780	0·11952		0·8451		**0·700**	5·012	

TABLE 9

x	x^2	$\sqrt{x}$	$\sqrt{10x}$	$\frac{1}{x}$	$\frac{1}{\sqrt{x}}$	$\frac{1}{\sqrt{10x}}$	log x	log t	t
7·00	49·0000	2·646	8·367 6	0·14286 21	0·3780	0·11952 8	0·8451 6	0·700	5·012 11
·01	49·1401	·648	·373 6	·14265 20	·3777	·11944 9	·8457 6	·701	·023 12
·02	49·2804	·650	·379 6	·14245 20	·3774	·11935 8	·8463 7	·702	·035 12
·03	49·4209	·651	·385 5	·14225 20	·3772	·11927 9	·8470 6	·703	·047 11
·04	49·5616	·653	·390 6	·14205 21	·3769	·11918 8	·8476 6	·704	·058 12
7·05	49·7025	2·655	8·396 6	0·14184 20	0·3766	0·11910 9	0·8482 6	0·705	5·070 12
·06	49·8436	·657	·402 6	·14164 20	·3764	·11901 8	·8488 6	·706	·082 11
·07	49·9849	·659	·408 6	·14144 20	·3761	·11893 8	·8494 6	·707	·093 12
·08	50·1264	·661	·414 6	·14124 20	·3758	·11885 9	·8500 6	·708	·105 12
·09	50·2681	·663	·420 6	·14104 19	·3756	·11876 8	·8506 7	·709	·117 12
7·10	50·4100	2·665	8·426 6	0·14085 20	0·3753	0·11868 9	0·8513 6	0·710	5·129 11
·11	50·5521	·666	·432 6	·14065 20	·3750	·11859 8	·8519 6	·711	·140 12
·12	50·6944	·668	·438 6	·14045 20	·3748	·11851 8	·8525 6	·712	·152 12
·13	50·8369	·670	·444 6	·14025 19	·3745	·11843 8	·8531 6	·713	·164 12
·14	50·9796	·672	·450 6	·14006 20	·3742	·11835 9	·8537 6	·714	·176 12
7·15	51·1225	2·674	8·456 6	0·13986 20	0·3740	0·11826 8	0·8543 6	0·715	5·188 12
·16	51·2656	·676	·462 6	·13966 19	·3737	·11818 8	·8549 6	·716	·200 12
·17	51·4089	·678	·468 5	·13947 19	·3735	·11810 8	·8555 6	·717	·212 12
·18	51·5524	·680	·473 6	·13928 20	·3732	·11802 9	·8561 6	·718	·224 12
·19	51·6961	·681	·479 6	·13908 19	·3729	·11793 8	·8567 6	·719	·236 12
7·20	51·8400	2·683	8·485 6	0·13889 19	0·3727	0·11785 8	0·8573 6	0·720	5·248 12
·21	51·9841	·685	·491 6	·13870 20	·3724	·11777 8	·8579 6	·721	·260 12
·22	52·1284	·687	·497 6	·13850 19	·3722	·11769 8	·8585 6	·722	·272 12
·23	52·2729	·689	·503 6	·13831 19	·3719	·11761 8	·8591 6	·723	·284 13
·24	52·4176	·691	·509 6	·13812 19	·3716	·11753 9	·8597 6	·724	·297 12
7·25	52·5625	2·693	8·515 6	0·13793 19	0·3714	0·11744 8	0·8603 6	0·725	5·309 12
·26	52·7076	·694	·521 5	·13774 19	·3711	·11736 8	·8609 6	·726	·321 12
·27	52·8529	·696	·526 6	·13755 19	·3709	·11728 8	·8615 6	·727	·333 13
·28	52·9984	·698	·532 6	·13736 19	·3706	·11720 8	·8621 6	·728	·346 12
·29	53·1441	·700	·538 6	·13717 18	·3704	·11712 8	·8627 6	·729	·358 12
7·30	53·2900	2·702	8·544 6	0·13699 19	0·3701	0·11704 8	0·8633 6	0·730	5·370 13
·31	53·4361	·704	·550 6	·13680 19	·3699	·11696 8	·8639 6	·731	·383 12
·32	53·5824	·706	·556 6	·13661 18	·3696	·11688 8	·8645 6	·732	·395 13
·33	53·7289	·707	·562 5	·13643 19	·3694	·11680 8	·8651 6	·733	·408 12
·34	53·8756	·709	·567 6	·13624 19	·3691	·11672 8	·8657 6	·734	·420 13
7·35	54·0225	2·711	8·573 6	0·13605 18	0·3689	0·11664 8	0·8663 6	0·735	5·433 12
·36	54·1696	·713	·579 6	·13587 18	·3686	·11656 8	·8669 6	·736	·445 13
·37	54·3169	·715	·585 6	·13569 19	·3684	·11648 7	·8675 6	·737	·458 12
·38	54·4644	·717	·591 6	·13550 18	·3681	·11641 8	·8681 5	·738	·470 13
·39	54·6121	·718	·597 5	·13532 18	·3679	·11633 8	·8686 6	·739	·483 12
7·40	54·7600	2·720	8·602 6	0·13514 19	0·3676	0·11625 8	0·8692 6	0·740	5·495 13
·41	54·9081	·722	·608 6	·13495 18	·3674	·11617 8	·8698 6	·741	·508 13
·42	55·0564	·724	·614 6	·13477 18	·3671	·11609 8	·8704 6	·742	·521 13
·43	55·2049	·726	·620 6	·13459 18	·3669	·11601 8	·8710 6	·743	·534 12
·44	55·3536	·728	·626 5	·13441 18	·3666	·11593 7	·8716 6	·744	·546 13
7·45	55·5025	2·729	8·631 6	0·13423 18	0·3664	0·11586 8	0·8722 5	0·745	5·559 13
·46	55·6516	·731	·637 6	·13405 18	·3661	·11578 8	·8727 6	·746	·572 13
·47	55·8009	·733	·643 6	·13387 18	·3659	·11570 8	·8733 6	·747	·585 13
·48	55·9504	·735	·649 5	·13369 18	·3656	·11562 7	·8739 6	·748	·598 12
·49	56·1001	·737	·654 6	·13351 18	·3654	·11555 8	·8745 6	·749	·610 13
7·50	56·2500	2·739	8·660	0·13333	0·3651	0·11547	0·8751	0·750	5·623

TABLE 9

x	x^2	$\sqrt{x}$	$\sqrt{10x}$		$\frac{1}{x}$		$\frac{1}{\sqrt{x}}$	$\frac{1}{\sqrt{10x}}$		$\log x$		$\log t$	t	
7·50	56·2500	2·739	8·660	6	0·13333	17	0·3651	0·11547	8	0·8751	5	**0·750**	5·623	13
·51	56·4001	·740	·666	6	·13316	18	·3649	·11539	7	·8756	6	**·751**	·636	13
·52	56·5504	·742	·672	6	·13298	18	·3647	·11532	8	·8762	6	**·752**	·649	13
·53	56·7009	·744	·678	5	·13280	17	·3644	·11524	8	·8768	6	**·753**	·662	13
·54	56·8516	·746	·683	6	·13263	18	·3642	·11516	7	·8774	5	**·754**	·675	14
7·55	57·0025	2·748	8·689	6	0·13245	17	0·3639	0·11509	8	0·8779	6	**0·755**	5·689	13
·56	57·1536	·750	·695	6	·13228	18	·3637	·11501	8	·8785	6	**·756**	·702	13
·57	57·3049	·751	·701	5	·13210	17	·3635	·11493	7	·8791	6	**·757**	·715	13
·58	57·4564	·753	·706	6	·13193	18	·3632	·11486	8	·8797	5	**·758**	·728	13
·59	57·6081	·755	·712	6	·13175	17	·3630	·11478	7	·8802	6	**·759**	·741	13
7·60	57·7600	2·757	8·718	6	0·13158	17	0·3627	0·11471	8	0·8808	6	**0·760**	5·754	14
·61	57·9121	·759	·724	5	·13141	18	·3625	·11463	7	·8814	6	**·761**	·768	13
·62	58·0644	·760	·729	6	·13123	17	·3623	·11456	8	·8820	5	**·762**	·781	13
·63	58·2169	·762	·735	6	·13106	17	·3620	·11448	7	·8825	6	**·763**	·794	14
·64	58·3696	·764	·741	5	·13089	17	·3618	·11441	8	·8831	6	**·764**	·808	13
7·65	58·5225	2·766	8·746	6	0·13072	17	0·3616	0·11433	7	0·8837	5	**0·765**	5·821	13
·66	58·6756	·768	·752	6	·13055	17	·3613	·11426	8	·8842	6	**·766**	·834	14
·67	58·8289	·769	·758	6	·13038	17	·3611	·11418	7	·8848	6	**·767**	·848	13
·68	58·9824	·771	·764	5	·13021	17	·3608	·11411	8	·8854	5	**·768**	·861	14
·69	59·1361	·773	·769	6	·13004	17	·3606	·11403	7	·8859	6	**·769**	·875	13
7·70	59·2900	2·775	8·775	6	0·12987	17	0·3604	0·11396	7	0·8865	6	**0·770**	5·888	14
·71	59·4441	·777	·781	5	·12970	17	·3601	·11389	8	·8871	5	**·771**	·902	14
·72	59·5984	·778	·786	6	·12953	16	·3599	·11381	7	·8876	6	**·772**	·916	13
·73	59·7529	·780	·792	6	·12937	17	·3597	·11374	7	·8882	5	**·773**	·929	14
·74	59·9076	·782	·798	5	·12920	17	·3594	·11367	8	·8887	6	**·774**	·943	14
7·75	60·0625	2·784	8·803	6	0·12903	16	0·3592	0·11359	7	0·8893	6	**0·775**	5·957	13
·76	60·2176	·786	·809	6	·12887	17	·3590	·11352	7	·8899	5	**·776**	·970	14
·77	60·3729	·787	·815	5	·12870	17	·3587	·11345	8	·8904	6	**·777**	·984	14
·78	60·5284	·789	·820	6	·12853	16	·3585	·11337	7	·8910	5	**·778**	5·998	14
·79	60·6841	·791	·826	6	·12837	16	·3583	·11330	7	·8915	6	**·779**	6·012	14
7·80	60·8400	2·793	8·832	5	0·12821	17	0·3581	0·11323	7	0·8921	6	**0·780**	6·026	13
·81	60·9961	·795	·837	6	·12804	16	·3578	·11316	8	·8927	5	**·781**	·039	14
·82	61·1524	·796	·843	6	·12788	17	·3576	·11308	7	·8932	6	**·782**	·053	14
·83	61·3089	·798	·849	5	·12771	16	·3574	·11301	7	·8938	5	**·783**	·067	14
·84	61·4656	·800	·854	6	·12755	16	·3571	·11294	7	·8943	6	**·784**	·081	14
7·85	61·6225	2·802	8·860	6	0·12739	16	0·3569	0·11287	8	0·8949	5	**0·785**	6·095	14
·86	61·7796	·804	·866	5	·12723	17	·3567	·11279	7	·8954	6	**·786**	·109	15
·87	61·9369	·805	·871	6	·12706	16	·3565	·11272	7	·8960	5	**·787**	·124	14
·88	62·0944	·807	·877	6	·12690	16	·3562	·11265	7	·8965	6	**·788**	·138	14
·89	62·2521	·809	·883	5	·12674	16	·3560	·11258	7	·8971	5	**·789**	·152	14
7·90	62·4100	2·811	8·888	6	0·12658	16	0·3558	0·11251	7	0·8976	6	**0·790**	6·166	14
·91	62·5681	·812	·894	5	·12642	16	·3556	·11244	7	·8982	5	**·791**	·180	14
·92	62·7264	·814	·899	6	·12626	16	·3553	·11237	7	·8987	6	**·792**	·194	15
·93	62·8849	·816	·905	6	·12610	16	·3551	·11230	7	·8993	5	**·793**	·209	14
·94	63·0436	·818	·911	5	·12594	15	·3549	·11223	8	·8998	6	**·794**	·223	14
7·95	63·2025	2·820	8·916	6	0·12579	16	0·3547	0·11215	7	0·9004	5	**0·795**	6·237	15
·96	63·3616	·821	·922	5	·12563	16	·3544	·11208	7	·9009	6	**·796**	·252	14
·97	63·5209	·823	·927	6	·12547	16	·3542	·11201	7	·9015	5	**·797**	·266	15
·98	63·6804	·825	·933	6	·12531	15	·3540	·11194	7	·9020	5	**·798**	·281	14
·99	63·8401	·827	·939	5	·12516	16	·3538	·11187	7	·9025	6	**·799**	·295	15
8·00	64·0000	2·828	8·944		0·12500		0·3536	0·11180		0·9031		**0·800**	6·310	

TABLE 9

x	x^2	$\sqrt{x}$	$\sqrt{10x}$		$\frac{1}{x}$		$\frac{1}{\sqrt{x}}$	$\frac{1}{\sqrt{10x}}$		$\log x$		$\log t$	t	
8·00	64·0000	**2·828**	8·944	6	0·12500	16	0·3536	0·11180	7	0·9031	5	**0·800**	6·310	14
·01	64·1601	·830	·950	5	·12484	15	·3533	·11173	7	·9036	6	**·801**	·324	15
·02	64·3204	·832	·955	6	·12469	16	·3531	·11166	7	·9042	5	**·802**	·339	14
·03	64·4809	·834	·961	6	·12453	15	·3529	·11159	7	·9047	6	**·803**	·353	15
·04	64·6416	·835	·967	5	·12438	16	·3527	·11152	6	·9053	5	**·804**	·368	15
8·05	64·8025	**2·837**	8·972	6	0·12422	15	0·3525	0·11146	7	0·9058	5	**0·805**	6·383	14
·06	64·9636	·839	·978	5	·12407	15	·3522	·11139	7	·9063	6	**·806**	·397	15
·07	65·1249	·841	·983	6	·12392	16	·3520	·11132	7	·9069	5	**·807**	·412	15
·08	65·2864	·843	·989	5	·12376	15	·3518	·11125	7	·9074	5	**·808**	·427	15
·09	65·4481	·844	8·994	6	·12361	15	·3516	·11118	7	·9079	6	**·809**	·442	15
8·10	65·6100	**2·846**	9·000	6	0·12346	16	0·3514	0·11111	7	0·9085	5	**0·810**	6·457	14
·11	65·7721	·848	·006	5	·12330	15	·3511	·11104	7	·9090	6	**·811**	·471	15
·12	65·9344	·850	·011	6	·12315	15	·3509	·11097	6	·9096	5	**·812**	·486	15
·13	66·0969	·851	·017	5	·12300	15	·3507	·11091	7	·9101	5	**·813**	·501	15
·14	66·2596	·853	·022	6	·12285	15	·3505	·11084	7	·9106	6	**·814**	·516	15
8·15	66·4225	**2·855**	9·028	5	0·12270	15	0·3503	0·11077	7	0·9112	5	**0·815**	6·531	15
·16	66·5856	·857	·033	6	·12255	15	·3501	·11070	7	·9117	5	**·816**	·546	15
·17	66·7489	·858	·039	5	·12240	15	·3499	·11063	6	·9122	6	**·817**	·561	16
·18	66·9124	·860	·044	6	·12225	15	·3496	·11057	7	·9128	5	**·818**	·577	15
·19	67·0761	·862	·050	5	·12210	15	·3494	·11050	7	·9133	5	**·819**	·592	15
8·20	67·2400	**2·864**	9·055	6	0·12195	15	0·3492	0·11043	7	0·9138	5	**0·820**	6·607	15
·21	67·4041	·865	·061	5	·12180	15	·3490	·11036	6	·9143	6	**·821**	·622	15
·22	67·5684	·867	·066	6	·12165	14	·3488	·11030	7	·9149	5	**·822**	·637	16
·23	67·7329	·869	·072	5	·12151	15	·3486	·11023	7	·9154	5	**·823**	·653	15
·24	67·8976	·871	·077	6	·12136	15	·3484	·11016	6	·9159	6	**·824**	·668	15
8·25	68·0625	**2·872**	9·083	5	0·12121	14	0·3482	0·11010	7	0·9165	5	**0·825**	6·683	16
·26	68·2276	·874	·088	6	·12107	15	·3479	·11003	7	·9170	5	**·826**	·699	15
·27	68·3929	·876	·094	5	·12092	15	·3477	·10996	6	·9175	5	**·827**	·714	16
·28	68·5584	·877	·099	6	·12077	14	·3475	·10990	7	·9180	6	**·828**	·730	15
·29	68·7241	·879	·105	5	·12063	15	·3473	·10983	7	·9186	5	**·829**	·745	16
8·30	68·8900	**2·881**	9·110	6	0·12048	14	0·3471	0·10976	6	0·9191	5	**0·830**	6·761	15
·31	69·0561	·883	·116	5	·12034	15	·3469	·10970	7	·9196	5	**·831**	·776	16
·32	69·2224	·884	·121	6	·12019	14	·3467	·10963	6	·9201	5	**·832**	·792	16
·33	69·3889	·886	·127	5	·12005	15	·3465	·10957	7	·9206	6	**·833**	·808	15
·34	69·5556	·888	·132	6	·11990	14	·3463	·10950	6	·9212	5	**·834**	·823	16
8·35	69·7225	**2·890**	9·138	5	0·11976	14	0·3461	0·10944	7	0·9217	5	**0·835**	6·839	16
·36	69·8896	·891	·143	6	·11962	15	·3459	·10937	7	·9222	5	**·836**	·855	16
·37	70·0569	·893	·149	5	·11947	14	·3457	·10930	6	·9227	5	**·837**	·871	16
·38	70·2244	·895	·154	6	·11933	14	·3454	·10924	7	·9232	6	**·838**	·887	15
·39	70·3921	·897	·160	5	·11919	14	·3452	·10917	6	·9238	5	**·839**	·902	16
8·40	70·5600	**2·898**	9·165	6	0·11905	14	0·3450	0·10911	7	0·9243	5	**0·840**	6·918	16
·41	70·7281	·900	·171	5	·11891	15	·3448	·10904	6	·9248	5	**·841**	·934	16
·42	70·8964	·902	·176	6	·11876	14	·3446	·10898	7	·9253	5	**·842**	·950	16
·43	71·0649	·903	·182	5	·11862	14	·3444	·10891	6	·9258	5	**·843**	·966	16
·44	71·2336	·905	·187	5	·11848	14	·3442	·10885	6	·9263	6	**·844**	·982	16
8·45	71·4025	**2·907**	9·192	6	0·11834	14	0·3440	0·10879	7	0·9269	5	**0·845**	6·998	17
·46	71·5716	·909	·198	5	·11820	14	·3438	·10872	6	·9274	5	**·846**	7·015	16
·47	71·7409	·910	·203	6	·11806	14	·3436	·10866	7	·9279	5	**·847**	·031	16
·48	71·9104	·912	·209	5	·11792	13	·3434	·10859	6	·9284	5	**·848**	·047	16
·49	72·0801	·914	·214	6	·11779	14	·3432	·10853	6	·9289	5	**·849**	·063	16
8·50	72·2500	2·915	9·220		0·11765		0·3430	0·10847		0·9294		**0·850**	7·079	

TABLE 9

x	x^2	$\sqrt{x}$	$\sqrt{10x}$		$\frac{1}{x}$		$\frac{1}{\sqrt{x}}$	$\frac{1}{\sqrt{10x}}$		$\log x$		$\log t$	t	
8·50	72·2500	2·915	9·220	5	0·11765	14	0·3430	0·10847	7	0·9294	5	0·850	7·079	17
·51	72·4201	·917	·225	5	·11751	14	·3428	·10840	6	·9299	5	·851	·096	16
·52	72·5904	·919	·230	6	·11737	14	·3426	·10834	7	·9304	5	·852	·112	17
·53	72·7609	·921	·236	5	·11723	13	·3424	·10827	6	·9309	6	·853	·129	16
·54	72·9316	·922	·241	6	·11710	14	·3422	·10821	6	·9315	5	·854	·145	16
8·55	73·1025	2·924	9·247	5	0·11696	14	0·3420	0·10815	7	0·9320	5	0·855	7·161	17
·56	73·2736	·926	·252	5	·11682	13	·3418	·10808	6	·9325	5	·856	·178	16
·57	73·4449	·927	·257	6	·11669	14	·3416	·10802	6	·9330	5	·857	·194	17
·58	73·6164	·929	·263	5	·11655	14	·3414	·10796	6	·9335	5	·858	·211	17
·59	73·7881	·931	·268	6	·11641	13	·3412	·10790	7	·9340	5	·859	·228	16
8·60	73·9600	2·933	9·274	5	0·11628	14	0·3410	0·10783	6	0·9345	5	0·860	7·244	17
·61	74·1321	·934	·279	5	·11614	13	·3408	·10777	6	·9350	5	·861	·261	17
·62	74·3044	·936	·284	6	·11601	14	·3406	·10771	6	·9355	5	·862	·278	17
·63	74·4769	·938	·290	5	·11587	13	·3404	·10765	7	·9360	5	·863	·295	16
·64	74·6496	·939	·295	6	·11574	13	·3402	·10758	6	·9365	5	·864	·311	17
8·65	74·8225	2·941	9·301	5	0·11561	14	0·3400	0·10752	6	0·9370	5	0·865	7·328	17
·66	74·9956	·943	·306	5	·11547	13	·3398	·10746	6	·9375	5	·866	·345	17
·67	75·1689	·944	·311	6	·11534	13	·3396	·10740	7	·9380	5	·867	·362	17
·68	75·3424	·946	·317	5	·11521	14	·3394	·10733	6	·9385	5	·868	·379	17
·69	75·5161	·948	·322	5	·11507	13	·3392	·10727	6	·9390	5	·869	·396	17
8·70	75·6900	2·950	9·327	6	0·11494	13	0·3390	0·10721	6	0·9395	5	0·870	7·413	17
·71	75·8641	·951	·333	5	·11481	13	·3388	·10715	6	·9400	5	·871	·430	17
·72	76·0384	·953	·338	5	·11468	13	·3386	·10709	6	·9405	5	·872	·447	17
·73	76·2129	·955	·343	6	·11455	13	·3384	·10703	6	·9410	5	·873	·464	18
·74	76·3876	·956	·349	5	·11442	13	·3383	·10697	7	·9415	5	·874	·482	17
8·75	76·5625	2·958	9·354	5	0·11429	13	0·3381	0·10690	6	0·9420	5	0·875	7·499	17
·76	76·7376	·960	·359	6	·11416	13	·3379	·10684	6	·9425	5	·876	·516	18
·77	76·9129	·961	·365	5	·11403	13	·3377	·10678	6	·9430	5	·877	·534	17
·78	77·0884	·963	·370	5	·11390	13	·3375	·10672	6	·9435	5	·878	·551	17
·79	77·2641	·965	·375	6	·11377	13	·3373	·10666	6	·9440	5	·879	·568	18
8·80	77·4400	2·966	9·381	5	0·11364	13	0·3371	0·10660	6	0·9445	5	0·880	7·586	17
·81	77·6161	·968	·386	5	·11351	13	·3369	·10654	6	·9450	5	·881	·603	18
·82	77·7924	·970	·391	6	·11338	13	·3367	·10648	6	·9455	5	·882	·621	17
·83	77·9689	·972	·397	5	·11325	13	·3365	·10642	6	·9460	5	·883	·638	18
·84	78·1456	·973	·402	5	·11312	13	·3363	·10636	6	·9465	4	·884	·656	18
8·85	78·3225	2·975	9·407	6	0·11299	12	0·3361	0·10630	6	0·9469	5	0·885	7·674	17
·86	78·4996	·977	·413	5	·11287	13	·3360	·10624	6	·9474	5	·886	·691	18
·87	78·6769	·978	·418	5	·11274	13	·3358	·10618	6	·9479	5	·887	·709	18
·88	78·8544	·980	·423	6	·11261	12	·3356	·10612	6	·9484	5	·888	·727	18
·89	79·0321	·982	·429	5	·11249	13	·3354	·10606	6	·9489	5	·889	·745	17
8·90	79·2100	2·983	9·434	5	0·11236	13	0·3352	0·10600	6	0·9494	5	0·890	7·762	18
·91	79·3881	·985	·439	6	·11223	12	·3350	·10594	6	·9499	5	·891	·780	18
·92	79·5664	·987	·445	5	·11211	13	·3348	·10588	6	·9504	5	·892	·798	18
·93	79·7449	·988	·450	5	·11198	12	·3346	·10582	6	·9509	4	·893	·816	18
·94	79·9236	·990	·455	5	·11186	13	·3345	·10576	6	·9513	5	·894	·834	18
8·95	80·1025	2·992	9·460	6	0·11173	12	0·3343	0·10570	6	0·9518	5	0·895	7·852	18
·96	80·2816	·993	·466	5	·11161	13	·3341	·10564	5	·9523	5	·896	·870	19
·97	80·4609	·995	·471	5	·11148	12	·3339	·10559	6	·9528	5	·897	·889	18
·98	80·6404	·997	·476	6	·11136	13	·3337	·10553	6	·9533	5	·898	·907	18
·99	80·8201	2·998	·482	5	·11123	12	·3335	·10547	6	·9538	4	·899	·925	18
9·00	81·0000	3·000	9·487		0·11111		0·3333	0·10541		0·9542		0·900	7·943	

TABLE 9

x	x^2	$\sqrt{x}$	$\sqrt{10x}$	d	$\frac{1}{x}$	d	$\frac{1}{\sqrt{x}}$	$\frac{1}{\sqrt{10x}}$	d	log x	d	log t	t	d
9·00	81·0000	3·000	9·487	5	0·11111	12	0·3333	0·10541	6	0·9542	5	0·900	7·943	19
·01	81·1801	·002	·492	5	·11099	13	·3331	·10535	6	·9547	5	·901	·962	18
·02	81·3604	·003	·497	6	·11086	12	·3330	·10529	6	·9552	5	·902	·980	18
·03	81·5409	·005	·503	5	·11074	12	·3328	·10523	5	·9557	5	·903	7·998	19
·04	81·7216	·007	·508	5	·11062	12	·3326	·10518	6	·9562	4	·904	8·017	18
9·05	81·9025	3·008	9·513	5	0·11050	12	0·3324	0·10512	6	0·9566	5	0·905	8·035	19
·06	82·0836	·010	·518	6	·11038	13	·3322	·10506	6	·9571	5	·906	·054	18
·07	82·2649	·012	·524	5	·11025	12	·3320	·10500	6	·9576	5	·907	·072	19
·08	82·4464	·013	·529	5	·11013	12	·3319	·10494	5	·9581	5	·908	·091	19
·09	82·6281	·015	·534	5	·11001	12	·3317	·10489	6	·9586	4	·909	·110	18
9·10	82·8100	3·017	9·539	6	0·10989	12	0·3315	0·10483	6	0·9590	5	0·910	8·128	19
·11	82·9921	·018	·545	5	·10977	12	·3313	·10477	6	·9595	5	·911	·147	19
·12	83·1744	·020	·550	5	·10965	12	·3311	·10471	5	·9600	5	·912	·166	19
·13	83·3569	·022	·555	5	·10953	12	·3310	·10466	6	·9605	4	·913	·185	19
·14	83·5396	·023	·560	6	·10941	12	·3308	·10460	6	·9609	5	·914	·204	18
9·15	83·7225	3·025	9·566	5	0·10929	12	0·3306	0·10454	6	0·9614	5	0·915	8·222	19
·16	83·9056	·027	·571	5	·10917	12	·3304	·10448	5	·9619	5	·916	·241	19
·17	84·0889	·028	·576	5	·10905	12	·3302	·10443	6	·9624	4	·917	·260	19
·18	84·2724	·030	·581	5	·10893	12	·3300	·10437	6	·9628	5	·918	·279	20
·19	84·4561	·032	·586	6	·10881	11	·3299	·10431	5	·9633	5	·919	·299	19
9·20	84·6400	3·033	9·592	5	0·10870	12	0·3297	0·10426	6	0·9638	5	0·920	8·318	19
·21	84·8241	·035	·597	5	·10858	12	·3295	·10420	6	·9643	4	·921	·337	19
·22	85·0084	·036	·602	5	·10846	12	·3293	·10414	5	·9647	5	·922	·356	19
·23	85·1929	·038	·607	5	·10834	11	·3292	·10409	6	·9652	5	·923	·375	20
·24	85·3776	·040	·612	6	·10823	12	·3290	·10403	5	·9657	4	·924	·395	19
9·25	85·5625	3·041	9·618	5	0·10811	12	0·3288	0·10398	6	0·9661	5	0·925	8·414	19
·26	85·7476	·043	·623	5	·10799	12	·3286	·10392	6	·9666	5	·926	·433	20
·27	85·9329	·045	·628	5	·10787	11	·3284	·10386	5	·9671	4	·927	·453	19
·28	86·1184	·046	·633	5	·10776	12	·3283	·10381	6	·9675	5	·928	·472	20
·29	86·3041	·048	·638	6	·10764	11	·3281	·10375	5	·9680	5	·929	·492	19
9·30	86·4900	3·050	9·644	5	0·10753	12	0·3279	0·10370	6	0·9685	4	0·930	8·511	20
·31	86·6761	·051	·649	5	·10741	11	·3277	·10364	6	·9689	5	·931	·531	20
·32	86·8624	·053	·654	5	·10730	12	·3276	·10358	5	·9694	5	·932	·551	19
·33	87·0489	·055	·659	5	·10718	11	·3274	·10353	6	·9699	4	·933	·570	20
·34	87·2356	·056	·664	6	·10707	12	·3272	·10347	5	·9703	5	·934	·590	20
9·35	87·4225	3·058	9·670	5	0·10695	11	0·3270	0·10342	6	0·9708	5	0·935	8·610	20
·36	87·6096	·059	·675	5	·10684	12	·3269	·10336	5	·9713	4	·936	·630	20
·37	87·7969	·061	·680	5	·10672	11	·3267	·10331	6	·9717	5	·937	·650	20
·38	87·9844	·063	·685	5	·10661	11	·3265	·10325	5	·9722	5	·938	·670	20
·39	88·1721	·064	·690	5	·10650	12	·3263	·10320	6	·9727	4	·939	·690	20
9·40	88·3600	3·066	9·695	6	0·10638	11	0·3262	0·10314	5	0·9731	5	0·940	8·710	20
·41	88·5481	·068	·701	5	·10627	11	·3260	·10309	6	·9736	5	·941	·730	20
·42	88·7364	·069	·706	5	·10616	12	·3258	·10303	5	·9741	4	·942	·750	20
·43	88·9249	·071	·711	5	·10604	11	·3256	·10298	6	·9745	5	·943	·770	20
·44	89·1136	·072	·716	5	·10593	11	·3255	·10292	5	·9750	4	·944	·790	20
9·45	89·3025	3·074	9·721	5	0·10582	11	0·3253	0·10287	6	0·9754	5	0·945	8·810	21
·46	89·4916	·076	·726	5	·10571	11	·3251	·10281	5	·9759	4	·946	·831	20
·47	89·6809	·077	·731	6	·10560	11	·3250	·10276	5	·9763	5	·947	·851	21
·48	89·8704	·079	·737	5	·10549	12	·3248	·10271	6	·9768	5	·948	·872	20
·49	90·0601	·081	·742	5	·10537	11	·3246	·10265	5	·9773	4	·949	·892	21
9·50	90·2500	3·082	9·747		0·10526		0·3244	0·10260		0·9777		0·950	8·913	

TABLE 9

x	x^2	$\sqrt{x}$	$\sqrt{10x}$	Δ	$\frac{1}{x}$	Δ	$\frac{1}{\sqrt{x}}$	$\frac{1}{\sqrt{10x}}$	Δ	$\log x$	Δ	$\log t$	t	Δ
9·50	90·2500	3·082	9·747	5	0·10526	11	0·3244	0·10260	6	0·9777	5	0·950	8·913	20
·51	90·4401	·084	·752	5	·10515	11	·3243	·10254	5	·9782	4	·951	·933	21
·52	90·6304	·085	·757	5	·10504	11	·3241	·10249	5	·9786	5	·952	·954	20
·53	90·8209	·087	·762	5	·10493	11	·3239	·10244	6	·9791	4	·953	·974	21
·54	91·0116	·089	·767	5	·10482	11	·3238	·10238	5	·9795	5	·954	8·995	21
9·55	91·2025	3·090	9·772	6	0·10471	11	0·3236	0·10233	5	0·9800	5	0·955	9·016	20
·56	91·3936	·092	·778	5	·10460	11	·3234	·10228	6	·9805	4	·956	·036	21
·57	91·5849	·094	·783	5	·10449	11	·3233	·10222	5	·9809	5	·957	·057	21
·58	91·7764	·095	·788	5	·10438	10	·3231	·10217	5	·9814	4	·958	·078	21
·59	91·9681	·097	·793	5	·10428	11	·3229	·10212	6	·9818	5	·959	·099	21
9·60	92·1600	3·098	9·798	5	0·10417	11	0·3227	0·10206	5	0·9823	4	0·960	9·120	21
·61	92·3521	·100	·803	5	·10406	11	·3226	·10201	5	·9827	5	·961	·141	21
·62	92·5444	·102	·808	5	·10395	11	·3224	·10196	6	·9832	4	·962	·162	21
·63	92·7369	·103	·813	5	·10384	11	·3222	·10190	5	·9836	5	·963	·183	21
·64	92·9296	·105	·818	5	·10373	10	·3221	·10185	5	·9841	4	·964	·204	22
9·65	93·1225	3·106	9·823	6	0·10363	11	0·3219	0·10180	6	0·9845	5	0·965	9·226	21
·66	93·3156	·108	·829	5	·10352	11	·3217	·10174	5	·9850	4	·966	·247	21
·67	93·5089	·110	·834	5	·10341	10	·3216	·10169	5	·9854	5	·967	·268	22
·68	93·7024	·111	·839	5	·10331	11	·3214	·10164	5	·9859	4	·968	·290	21
·69	93·8961	·113	·844	5	·10320	11	·3212	·10159	6	·9863	5	·969	·311	22
9·70	94·0900	3·114	9·849	5	0·10309	10	0·3211	0·10153	5	0·9868	4	0·970	9·333	21
·71	94·2841	·116	·854	5	·10299	11	·3209	·10148	5	·9872	5	·971	·354	22
·72	94·4784	·118	·859	5	·10288	11	·3208	·10143	5	·9877	4	·972	·376	21
·73	94·6729	·119	·864	5	·10277	10	·3206	·10138	5	·9881	5	·973	·397	22
·74	94·8676	·121	·869	5	·10267	11	·3204	·10133	6	·9886	4	·974	·419	22
9·75	95·0625	3·122	9·874	5	0·10256	10	0·3203	0·10127	5	0·9890	4	0·975	9·441	21
·76	95·2576	·124	·879	5	·10246	11	·3201	·10122	5	·9894	5	·976	·462	22
·77	95·4529	·126	·884	5	·10235	10	·3199	·10117	5	·9899	4	·977	·484	22
·78	95·6484	·127	·889	5	·10225	10	·3198	·10112	5	·9903	5	·978	·506	22
·79	95·8441	·129	·894	5	·10215	11	·3196	·10107	5	·9908	4	·979	·528	22
9·80	96·0400	3·130	9·899	6	0·10204	10	0·3194	0·10102	6	0·9912	5	0·980	9·550	22
·81	96·2361	·132	·905	5	·10194	11	·3193	·10096	5	·9917	4	·981	·572	22
·82	96·4324	·134	·910	5	·10183	10	·3191	·10091	5	·9921	5	·982	·594	22
·83	96·6289	·135	·915	5	·10173	10	·3190	·10086	5	·9926	4	·983	·616	22
·84	96·8256	·137	·920	5	·10163	11	·3188	·10081	5	·9930	4	·984	·638	23
9·85	97·0225	3·138	9·925	5	0·10152	10	0·3186	0·10076	5	0·9934	5	0·985	9·661	22
·86	97·2196	·140	·930	5	·10142	10	·3185	·10071	5	·9939	4	·986	·683	22
·87	97·4169	·142	·935	5	·10132	11	·3183	·10066	5	·9943	5	·987	·705	22
·88	97·6144	·143	·940	5	·10121	10	·3181	·10061	6	·9948	4	·988	·727	23
·89	97·8121	·145	·945	5	·10111	10	·3180	·10055	5	·9952	4	·989	·750	22
9·90	98·0100	3·146	9·950	5	0·10101	10	0·3178	0·10050	5	0·9956	5	0·990	9·772	23
·91	98·2081	·148	·955	5	·10091	10	·3177	·10045	5	·9961	4	·991	·795	22
·92	98·4064	·150	·960	5	·10081	11	·3175	·10040	5	·9965	4	·992	·817	23
·93	98·6049	·151	·965	5	·10070	10	·3173	·10035	5	·9969	5	·993	·840	23
·94	98·8036	·153	·970	5	·10060	10	·3172	·10030	5	·9974	4	·994	·863	23
9·95	99·0025	3·154	9·975	5	0·10050	10	0·3170	0·10025	5	0·9978	5	0·995	9·886	22
·96	99·2016	·156	·980	5	·10040	10	·3169	·10020	5	·9983	4	·996	·908	23
·97	99·4009	·158	·985	5	·10030	10	·3167	·10015	5	·9987	4	·997	·931	23
·98	99·6004	·159	·990	5	·10020	10	·3165	·10010	5	·9991	5	·998	·954	23
·99	99·8001	·161	9·995	5	·10010	10	·3164	·10005	5	0·9996	4	·999	9·977	23
10·00	100·0000	3·162	10·000		0·10000		0·3162	0·10000		1·0000		1·000	10·000	

TABLE 10. LOGARITHMS OF FACTORIALS

n	log *n*!	*n*	log *n*!	*n*	log *n*!	*n*	log *n*!	*n*	log *n*!	*n*	log *n*!
0	0·0000	**50**	64·4831	**100**	157·9700	**150**	262·7569	**200**	374·8969	**250**	492·5096
1	0·0000	**51**	66·1906	**101**	159·9743	**151**	264·9359	**201**	377·2001	**251**	494·9093
2	0·3010	**52**	67·9066	**102**	161·9829	**152**	267·1177	**202**	379·5054	**252**	497·3107
3	0·7782	**53**	69·6309	**103**	163·9958	**153**	269·3024	**203**	381·8129	**253**	499·7138
4	1·3802	**54**	71·3633	**104**	166·0128	**154**	271·4899	**204**	384·1226	**254**	502·1186
5	2·0792	**55**	73·1037	**105**	168·0340	**155**	273·6803	**205**	386·4343	**255**	504·5252
6	2·8573	**56**	74·8519	**106**	170·0593	**156**	275·8734	**206**	388·7482	**256**	506·9334
7	3·7024	**57**	76·6077	**107**	172·0887	**157**	278·0693	**207**	391·0642	**257**	509·3433
8	4·6055	**58**	78·3712	**108**	174·1221	**158**	280·2679	**208**	393·3822	**258**	511·7549
9	5·5598	**59**	80·1420	**109**	176·1595	**159**	282·4693	**209**	395·7024	**259**	514·1682
10	6·5598	**60**	81·9202	**110**	178·2009	**160**	284·6735	**210**	398·0246	**260**	516·5832
11	7·6012	**61**	83·7055	**111**	180·2462	**161**	286·8803	**211**	400·3489	**261**	518·9999
12	8·6803	**62**	85·4979	**112**	182·2955	**162**	289·0898	**212**	402·6752	**262**	521·4182
13	9·7943	**63**	87·2972	**113**	184·3485	**163**	291·3020	**213**	405·0036	**263**	523·8381
14	10·9404	**64**	89·1034	**114**	186·4054	**164**	293·5168	**214**	407·3340	**264**	526·2597
15	12·1165	**65**	90·9163	**115**	188·4661	**165**	295·7343	**215**	409·6664	**265**	528·6830
16	13·3206	**66**	92·7359	**116**	190·5306	**166**	297·9544	**216**	412·0009	**266**	531·1078
17	14·5511	**67**	94·5619	**117**	192·5988	**167**	300·1771	**217**	414·3373	**267**	533·5344
18	15·8063	**68**	96·3945	**118**	194·6707	**168**	302·4024	**218**	416·6758	**268**	535·9625
19	17·0851	**69**	98·2333	**119**	196·7462	**169**	304·6303	**219**	419·0162	**269**	538·3922
20	18·3861	**70**	100·0784	**120**	198·8254	**170**	306·8608	**220**	421·3587	**270**	540·8236
21	19·7083	**71**	101·9297	**121**	200·9082	**171**	309·0938	**221**	423·7031	**271**	543·2566
22	21·0508	**72**	103·7870	**122**	202·9945	**172**	311·3293	**222**	426·0494	**272**	545·6912
23	22·4125	**73**	105·6503	**123**	205·0844	**173**	313·5674	**223**	428·3977	**273**	548·1273
24	23·7927	**74**	107·5196	**124**	207·1779	**174**	315·8079	**224**	430·7480	**274**	550·5651
25	25·1906	**75**	109·3946	**125**	209·2748	**175**	318·0509	**225**	433·1002	**275**	553·0044
26	26·6056	**76**	111·2754	**126**	211·3751	**176**	320·2965	**226**	435·4543	**276**	555·4453
27	28·0370	**77**	113·1619	**127**	213·4790	**177**	322·5444	**227**	437·8103	**277**	557·8878
28	29·4841	**78**	115·0540	**128**	215·5862	**178**	324·7948	**228**	440·1682	**278**	560·3318
29	30·9465	**79**	116·9516	**129**	217·6967	**179**	327·0477	**229**	442·5281	**279**	562·7774
30	32·4237	**80**	118·8547	**130**	219·8107	**180**	329·3030	**230**	444·8898	**280**	565·2246
31	33·9150	**81**	120·7632	**131**	221·9280	**181**	331·5606	**231**	447·2534	**281**	567·6733
32	35·4202	**82**	122·6770	**132**	224·0485	**182**	333·8207	**232**	449·6189	**282**	570·1235
33	36·9387	**83**	124·5961	**133**	226·1724	**183**	336·0832	**233**	451·9862	**283**	572·5753
34	38·4702	**84**	126·5204	**134**	228·2995	**184**	338·3480	**234**	454·3555	**284**	575·0287
35	40·0142	**85**	128·4498	**135**	230·4298	**185**	340·6152	**235**	456·7265	**285**	577·4835
36	41·5705	**86**	130·3843	**136**	232·5634	**186**	342·8847	**236**	459·0994	**286**	579·9399
37	43·1387	**87**	132·3238	**137**	234·7001	**187**	345·1565	**237**	461·4742	**287**	582·3977
38	44·7185	**88**	134·2683	**138**	236·8400	**188**	347·4307	**238**	463·8508	**288**	584·8571
39	46·3096	**89**	136·2177	**139**	238·9830	**189**	349·7071	**239**	466·2292	**289**	587·3180
40	47·9116	**90**	138·1719	**140**	241·1291	**190**	351·9859	**240**	468·6094	**290**	589·7804
41	49·5244	**91**	140·1310	**141**	243·2783	**191**	354·2669	**241**	470·9914	**291**	592·2443
42	51·1477	**92**	142·0948	**142**	245·4306	**192**	356·5502	**242**	473·3752	**292**	594·7097
43	52·7811	**93**	144·0632	**143**	247·5860	**193**	358·8358	**243**	475·7608	**293**	597·1766
44	54·4246	**94**	146·0364	**144**	249·7443	**194**	361·1236	**244**	478·1482	**294**	599·6449
45	56·0778	**95**	148·0141	**145**	251·9057	**195**	363·4136	**245**	480·5374	**295**	602·1147
46	57·7406	**96**	149·9964	**146**	254·0700	**196**	365·7059	**246**	482·9283	**296**	604·5860
47	59·4127	**97**	151·9831	**147**	256·2374	**197**	368·0003	**247**	485·3210	**297**	607·0588
48	61·0939	**98**	153·9744	**148**	258·4076	**198**	370·2970	**248**	487·7154	**298**	609·5330
49	62·7841	**99**	155·9700	**149**	260·5808	**199**	372·5959	**249**	490·1116	**299**	612·0087
50	64·4831	**100**	157·9700	**150**	262·7569	**200**	374·8969	**250**	492·5096	**300**	614·4858

For large *n* $\log n! \doteqdot 0{\cdot}39909 + (n+\tfrac{1}{2}) \log n - 0{\cdot}43429\,45\, n$

PROPORTIONAL PARTS

	2	3	4	5	6	7	8	9	10	11	12	13	14	15	16	17	18	19
1	0·2	0·3	0·4	0·5	0·6	0·7	0·8	0·9	1·0	1·1	1·2	1·3	1·4	1·5	1·6	1·7	1·8	1·9
2	0·4	0·6	0·8	1·0	1·2	1·4	1·6	1·8	2·0	2·2	2·4	2·6	2·8	3·0	3·2	3·4	3·6	3·8
3	0·6	0·9	1·2	1·5	1·8	2·1	2·4	2·7	3·0	3·3	3·6	3·9	4·2	4·5	4·8	5·1	5·4	5·7
4	0·8	1·2	1·6	2·0	2·4	2·8	3·2	3·6	4·0	4·4	4·8	5·2	5·6	6·0	6·4	6·8	7·2	7·6
5	1·0	1·5	2·0	2·5	3·0	3·5	4·0	4·5	5·0	5·5	6·0	6·5	7·0	7·5	8·0	8·5	9·0	9·5
6	1·2	1·8	2·4	3·0	3·6	4·2	4·8	5·4	6·0	6·6	7·2	7·8	8·4	9·0	9·6	10·2	10·8	11·4
7	1·4	2·1	2·8	3·5	4·2	4·9	5·6	6·3	7·0	7·7	8·4	9·1	9·8	10·5	11·2	11·9	12·6	13·3
8	1·6	2·4	3·2	4·0	4·8	5·6	6·4	7·2	8·0	8·8	9·6	10·4	11·2	12·0	12·8	13·6	14·4	15·2
9	1·8	2·7	3·6	4·5	5·4	6·3	7·2	8·1	9·0	9·9	10·8	11·7	12·6	13·5	14·4	15·3	16·2	17·1

	20	21	22	23	24	25	26	27	28	29	30	31	32	33	34	35
1	2·0	2·1	2·2	2·3	2·4	2·5	2·6	2·7	2·8	2·9	3·0	3·1	3·2	3·3	3·4	3·5
2	4·0	4·2	4·4	4·6	4·8	5·0	5·2	5·4	5·6	5·8	6·0	6·2	6·4	6·6	6·8	7·0
3	6·0	6·3	6·6	6·9	7·2	7·5	7·8	8·1	8·4	8·7	9·0	9·3	9·6	9·9	10·2	10·5
4	8·0	8·4	8·8	9·2	9·6	10·0	10·4	10·8	11·2	11·6	12·0	12·4	12·8	13·2	13·6	14·0
5	10·0	10·5	11·0	11·5	12·0	12·5	13·0	13·5	14·0	14·5	15·0	15·5	16·0	16·5	17·0	17·5
6	12·0	12·6	13·2	13·8	14·4	15·0	15·6	16·2	16·8	17·4	18·0	18·6	19·2	19·8	20·4	21·0
7	14·0	14·7	15·4	16·1	16·8	17·5	18·2	18·9	19·6	20·3	21·0	21·7	22·4	23·1	23·8	24·5
8	16·0	16·8	17·6	18·4	19·2	20·0	20·8	21·6	22·4	23·2	24·0	24·8	25·6	26·4	27·2	28·0
9	18·0	18·9	19·8	20·7	21·6	22·5	23·4	24·3	25·2	26·1	27·0	27·9	28·8	29·7	30·6	31·5

	36	37	38	39	40	41	42	43	44	45	46	47	48	49	50	51
1	3·6	3·7	3·8	3·9	4·0	4·1	4·2	4·3	4·4	4·5	4·6	4·7	4·8	4·9	5·0	5·1
2	7·2	7·4	7·6	7·8	8·0	8·2	8·4	8·6	8·8	9·0	9·2	9·4	9·6	9·8	10·0	10·2
3	10·8	11·1	11·4	11·7	12·0	12·3	12·6	12·9	13·2	13·5	13·8	14·1	14·4	14·7	15·0	15·3
4	14·4	14·8	15·2	15·6	16·0	16·4	16·8	17·2	17·6	18·0	18·4	18·8	19·2	19·6	20·0	20·4
5	18·0	18·5	19·0	19·5	20·0	20·5	21·0	21·5	22·0	22·5	23·0	23·5	24·0	24·5	25·0	25·5
6	21·6	22·2	22·8	23·4	24·0	24·6	25·2	25·8	26·4	27·0	27·6	28·2	28·8	29·4	30·0	30·6
7	25·2	25·9	26·6	27·3	28·0	28·7	29·4	30·1	30·8	31·5	32·2	32·9	33·6	34·3	35·0	35·7
8	28·8	29·6	30·4	31·2	32·0	32·8	33·6	34·4	35·2	36·0	36·8	37·6	38·4	39·2	40·0	40·8
9	32·4	33·3	34·2	35·1	36·0	36·9	37·8	38·7	39·6	40·5	41·4	42·3	43·2	44·1	45·0	45·9

	52	53	54	55	56	57	58	59	60	61	62	63	64	65	66	67
1	5·2	5·3	5·4	5·5	5·6	5·7	5·8	5·9	6·0	6·1	6·2	6·3	6·4	6·5	6·6	6·7
2	10·4	10·6	10·8	11·0	11·2	11·4	11·6	11·8	12·0	12·2	12·4	12·6	12·8	13·0	13·2	13·4
3	15·6	15·9	16·2	16·5	16·8	17·1	17·4	17·7	18·0	18·3	18·6	18·9	19·2	19·5	19·8	20·1
4	20·8	21·2	21·6	22·0	22·4	22·8	23·2	23·6	24·0	24·4	24·8	25·2	25·6	26·0	26·4	26·8
5	26·0	26·5	27·0	27·5	28·0	28·5	29·0	29·5	30·0	30·5	31·0	31·5	32·0	32·5	33·0	33·5
6	31·2	31·8	32·4	33·0	33·6	34·2	34·8	35·4	36·0	36·6	37·2	37·8	38·4	39·0	39·6	40·2
7	36·4	37·1	37·8	38·5	39·2	39·9	40·6	41·3	42·0	42·7	43·4	44·1	44·8	45·5	46·2	46·9
8	41·6	42·4	43·2	44·0	44·8	45·6	46·4	47·2	48·0	48·8	49·6	50·4	51·2	52·0	52·8	53·6
9	46·8	47·7	48·6	49·5	50·4	51·3	52·2	53·1	54·0	54·9	55·8	56·7	57·6	58·5	59·4	60·3

	68	69	70	71	72	73	74	75	76	77	78	79	80	81	82	83
1	6·8	6·9	7·0	7·1	7·2	7·3	7·4	7·5	7·6	7·7	7·8	7·9	8·0	8·1	8·2	8·3
2	13·6	13·8	14·0	14·2	14·4	14·6	14·8	15·0	15·2	15·4	15·6	15·8	16·0	16·2	16·4	16·6
3	20·4	20·7	21·0	21·3	21·6	21·9	22·2	22·5	22·8	23·1	23·4	23·7	24·0	24·3	24·6	24·9
4	27·2	27·6	28·0	28·4	28·8	29·2	29·6	30·0	30·4	30·8	31·2	31·6	32·0	32·4	32·8	33·2
5	34·0	34·5	35·0	35·5	36·0	36·5	37·0	37·5	38·0	38·5	39·0	39·5	40·0	40·5	41·0	41·5
6	40·8	41·4	42·0	42·6	43·2	43·8	44·4	45·0	45·6	46·2	46·8	47·4	48·0	48·6	49·2	49·8
7	47·6	48·3	49·0	49·7	50·4	51·1	51·8	52·5	53·2	53·9	54·6	55·3	56·0	56·7	57·4	58·1
8	54·4	55·2	56·0	56·8	57·6	58·4	59·2	60·0	60·8	61·6	62·4	63·2	64·0	64·8	65·6	66·4
9	61·2	62·1	63·0	63·9	64·8	65·7	66·6	67·5	68·4	69·3	70·2	71·1	72·0	72·9	73·8	74·7

	84	85	86	87	88	89	90	91	92	93	94	95	96	97	98	99
1	8·4	8·5	8·6	8·7	8·8	8·9	9·0	9·1	9·2	9·3	9·4	9·5	9·6	9·7	9·8	9·9
2	16·8	17·0	17·2	17·4	17·6	17·8	18·0	18·2	18·4	18·6	18·8	19·0	19·2	19·4	19·6	19·8
3	25·2	25·5	25·8	26·1	26·4	26·7	27·0	27·3	27·6	27·9	28·2	28·5	28·8	29·1	29·4	29·7
4	33·6	34·0	34·4	34·8	35·2	35·6	36·0	36·4	36·8	37·2	37·6	38·0	38·4	38·8	39·2	39·6
5	42·0	42·5	43·0	43·5	44·0	44·5	45·0	45·5	46·0	46·5	47·0	47·5	48·0	48·5	49·0	49·5
6	50·4	51·0	51·6	52·2	52·8	53·4	54·0	54·6	55·2	55·8	56·4	57·0	57·6	58·2	58·8	59·4
7	58·8	59·5	60·2	60·9	61·6	62·3	63·0	63·7	64·4	65·1	65·8	66·5	67·2	67·9	68·6	69·3
8	67·2	68·0	68·8	69·6	70·4	71·2	72·0	72·8	73·6	74·4	75·2	76·0	76·8	77·6	78·4	79·2
9	75·6	76·5	77·4	78·3	79·2	80·1	81·0	81·9	82·8	83·7	84·6	85·5	86·4	87·3	88·2	89·1

Published by the Syndics of the Cambridge University Press
The Pitt Building, Trumpington Street, Cambridge CB2 1RP
Bentley House, 200 Euston Road, London NW1 2DB
32 East 57th Street, New York, NY 10022, USA
296 Beaconsfield Parade, Middle Park, Melbourne 3206, Australia

ISBN 0 521 05564 4

First published 1953
Reprinted 1958 1961 1962 1964
1966 1968 1970 1971 1973 1974 1975 1978

Printed in Great Britain at the
University Press, Cambridge